ELECTRONIC COMMUNICATIONS

A Step-by-Step Introduction

Heathkit/Zenith Educational Systems

A SPECTRUM BOOK

Prentice-Hall, Inc. Englewood Cliffs, New Jersey 07632

Library of Congress Cataloging in Publication Data

Main entry under title:

Electronic communications.

"A Spectrum Book."

Reproduces the text of the ed. published: Benton Harbor, Mich. : Heath Co., ©1981; unit objectives and self-reviews omitted.

Includes index.

1. Radio. I. Heathkit/Zenith Educational Systems (Group)

TK6550.E36 1983 621.3841 82-23158
ISBN 0-13-250423-5
ISBN 0-13-250415-4 (pbk.)

A SPECTRUM BOOK

10 9 8 7 6 5 4 3 2 1

Printed in the United States of America

Manufacturing buyer Patrick Mahoney

ISBN 0-13-250423-5

ISBN 0-13-250415-4 {PBK.}

Prentice-Hall International, Inc., *London*
Prentice-Hall of Australia Pty. Limited, *Sydney*
Prentice-Hall of Canada Inc., *Toronto*
Prentice-Hall of India Private Limited, *New Delhi*
Prentice-Hall of Japan, Inc., *Tokyo*
Prentice-Hall of Southeast Asia Pte. Ltd., *Singapore*
Whitehall Books Limited, *Wellington, New Zealand*
Editora Prentice-Hall do Brasil Ltda., *Rio de Janeiro*

CONTENTS

INTRODUCTION

Throughout time, the one thing that has distinguished humans from other animals is their ability to exchange ideas and other information. That is, humans can **communicate**. It is this capability that has played perhaps the biggest part in the development of civilization. In fact, as our civilization continues to grow, so must our communications capacity. The two are intertwined.

It comes as no surprise then that one of the first applications of the "new" field of electricity was to extend our communications range. This was accomplished through the use of wires and telegraphy. Messages were sent by turning electrical currents on and off in accordance with a telegraph code. This system gradually evolved into the telephone system where the electrical currents are varied at an audio rate. Thus, the spoken word can be conveyed between two distant points. However, the telephone system still required wires, which limited its capabilities. Thus, the next development was to be "wireless" communications in the form of radio. This greatly extended the communications range and, for the first time, included ships at sea and remote areas of the world. Actually, it allowed instantaneous and virtually world-wide communications for the first time.

"Wireless" or radio communications represented a significant advancement. And, as a result, a whole branch of electronics was established. This course is intended to teach you the fundamental and advanced techniques of that branch: electronic communications.

Unit 1

COMMUNICATIONS FUNDAMENTALS

INTRODUCTION

This unit discusses the fundamentals of communications. It examines, in detail, communications systems, information, transmitters, receivers, noise, and, more importantly, modulation. A thorough understanding of modulation is of the utmost importance, because without it, radio communications would be impossible. For this reason, a large portion of this unit is devoted to the discussion of the three different types of modulation.

WHAT IS COMMUNICATION?

Communication is defined as "a process by which information is exchanged." In electronics, it is the transmission and reception of information. Likewise, information is defined as "the communication of knowledge or intelligence." For the purposes of this course, it is defined as any electrical signal representing data. Thus, the purpose of any communications system is to convey or transfer information from one point to another.

Information Transfer

Our most fundamental methods of transferring information are speech and, as in this case, writing. Communication of the written word developed from hand-carried letters and newspapers to the mail system, telegraph, and now electronic mail. Spoken communications evolved from face-to-face contact into telephone and radio communication. All of these steps were taken in an effort to increase the communications distance and speed.

The most significant advance in increasing communications range was radio. Basically, the audio or sound waves are converted to an electrical signal then into audio waves and transmitted to a distant receiving station. However, if the audio signal is transmitted at its original frequency, a number of problems present themselves. First to be efficient, the transmitting antenna must be at least 1/4 to 1/2 wavelength long. This means that for a 3,000 Hz signal, the antenna would have to be at least 25 kilometers long! Second, even if the antenna problem was solved, only one station could transmit at a time. This is because all stations would be operating on the same "audio" frequencies. And third, transmission systems at these frequencies are very inefficient.

All of these problems can be solved by using a higher frequency signal as a "carrier" for the audio information. In essence, the speech signal is transferred to a much higher frequency for transmission, then converted back to audio frequencies by the receiver. The former is called **modulation**, while the latter is **demodulation**.

Modulation

In the process of modulation, some characteristic of a high frequency sine wave is varied in accordance with the information or **modulating** signal. This signal may be an audio waveform, a digital pulse train, a television picture, or any other form of information. The important consideration is that it is transferred to a higher frequency for efficient transmission.

As mentioned before, the modulated high frequency sine wave is called the **carrier**. The mathematical expression for an unmodulated sine wave or carrier is:

$$e = A \sin(\omega t + \phi)$$

Where

e = instantaneous value of the wave (voltage or current)

A = maximum amplitude

ω = angular velocity ($2\pi f$)

t = time

ϕ = phase angle

This equation shows that there are three characteristics of the wave that can be varied or modulated. These are: amplitude (A), angular velocity or frequency (ω), and phase angle (ϕ).

Types of Modulation

Since, three characteristics of the sine wave carrier can be varied, it follows that there are three types of modulation. These are amplitude modulation (AM), frequency modulation (FM), and phase modulation (PM). However, in practice, it is very difficult to distinguish between phase and frequency modulation. Therefore, these two types of modulation are grouped together under the title of **Angle** modulation. Thus, there are two basic types of modulation: Amplitude and Angle. The next section will discuss both of these in detail.

REVIEW OF COMPLEX WAVES

In the first section, the need for modulation and its basic principles were established. Now, before continuing the study of modulation, a quick review of complex waves and frequency domain analysis is needed.

Frequency domain analysis stems from the fact that any periodic waveform is made up of sine waves. A periodic wave is one that has the same waveshape from one cycle to the next. All the waveforms shown in Figure 1-1 are periodic waves if they have the same shape on each and every cycle. Any waveform of this type can be formed by the superimposing of a number of waves which have certain amplitude, phase, and frequency characteristics. Similarly, regardless of how a particular periodic waveform is formed, it can be proven that it consists of sine waves in a specific phase, frequency, and amplitude relationship.

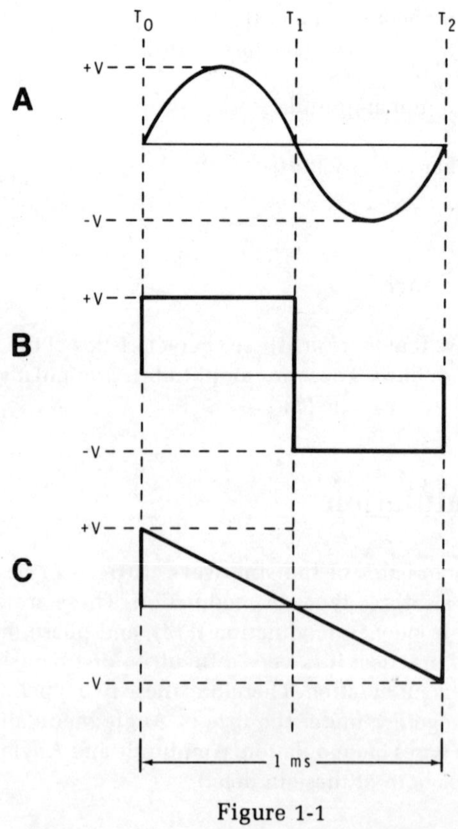

Figure 1-1
A sine wave (A), square wave (B), and sawtooth wave (C).

Sine Wave

Since the sine wave is the basis of all waveforms, let's examine it in more detail. Figure 1-1A is a **time domain** representation of the sine wave. It shows the wave as it would appear on an oscilloscope display, which is a graph of time versus amplitude.

Figure 1-2 shows the wave as it would appear on a spectrum analyzer display. This device analyzes the wave's **frequency domain**. As you can see, it gives us a graph of frequency versus amplitude. It is actually a bar graph showing all the frequency components contained within a wave. Since this sine wave has a period of 1 ms, its frequency is 1 kHz. Therefore, a bar representing the wave's amplitude appears at 1 kHz. There are no other frequency components, since this is a pure sine wave.

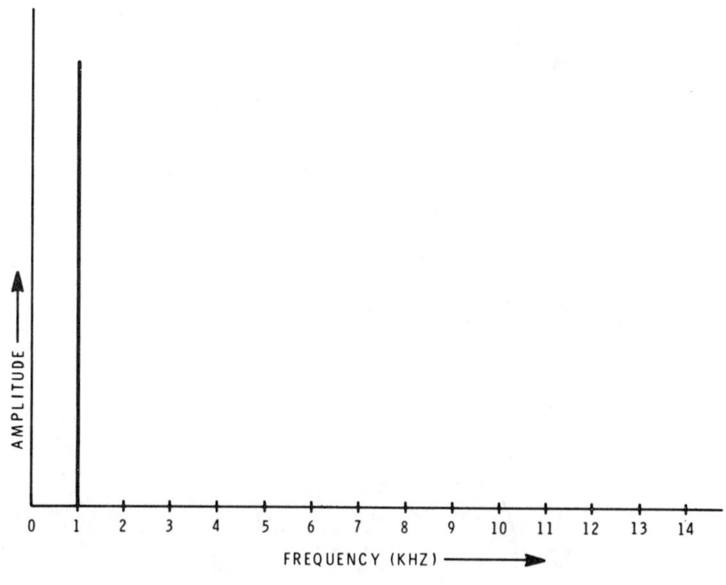

Figure 1-2
A spectrum analyzer display of the frequency domain.

Square Wave

Figure 1-3A shows one cycle of a 1 kHz square wave. Since this is a periodic wave, this cycle is repeated over and over again. This waveform is made up of a large number of sine waves. More specifically, a perfect square wave is composed of a fundamental frequency and an infinite number of odd harmonics.

The result of adding the first harmonic or fundamental and the third harmonic is shown in Figure 1-3B. The addition of the fifth harmonic is shown in Figure 1-3C. Note that the resultant waveform is already approaching a true square wave. In fact, as more odd-order harmonics are added, the closer the waveform gets to becoming a "perfect" square wave.

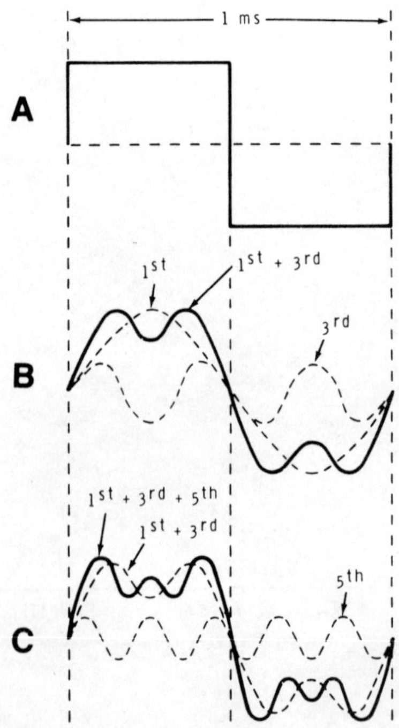

Figure 1-3
The composition of a square wave.

It should now be a simple matter to plot the frequency domain of a square wave since we know that it is composed of odd-order harmonics. Figure 1-4 shows the results. It illustrates the correct amplitude for all the harmonics out to 13 kHz. For a perfect square wave, the chart would extend to infinity.

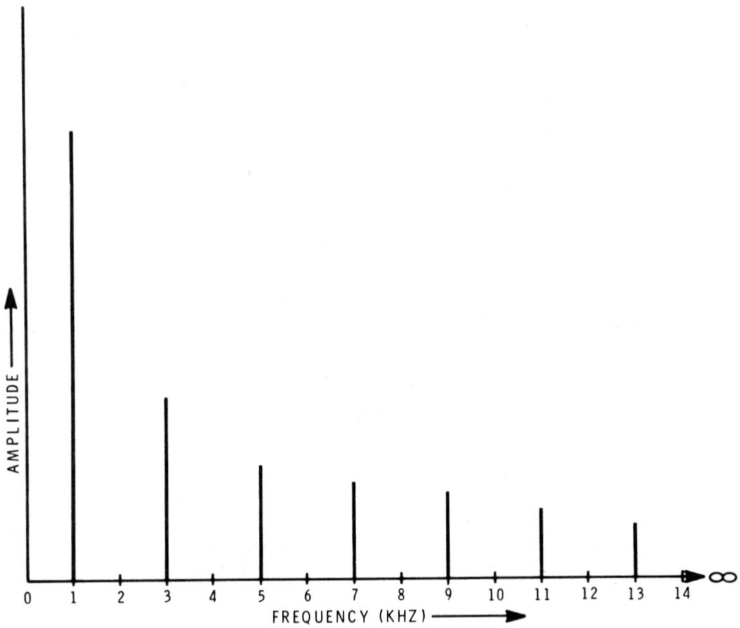

Figure 1-4
The frequency domain analysis of a square wave.

Sawtooth Wave

Figure 1-5 illustrates that a different type of waveform is developed when the harmonic content is changed. Here both even and odd-order harmonics are added. The result is a sawtooth wave. Figure 1-5C shows that, with just the first through fourth harmonics present, the resultant waveform comes very close to the perfect sawtooth.

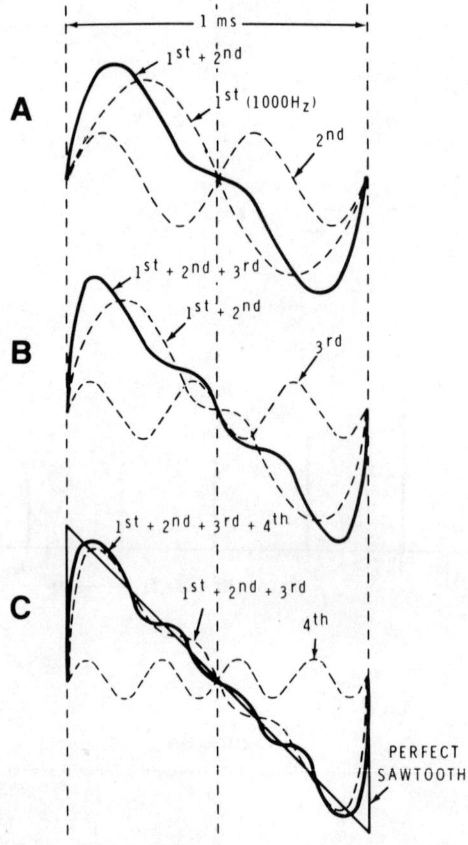

Figure 1-5
The superimposing of sine waves to form a sawtooth wave.

The frequency domain analysis of a sawtooth is shown in Figure 1-6. Note that all harmonics are present, both even and odd-ordered.

The sawtooth and square wave are but two examples of the unlimited variety of complex waveforms. The important concept is that sine waves can be combined in an infinite number of ways to produce an infinite variety of waveforms. By the same token, any waveform can be broken down into its sine wave components.

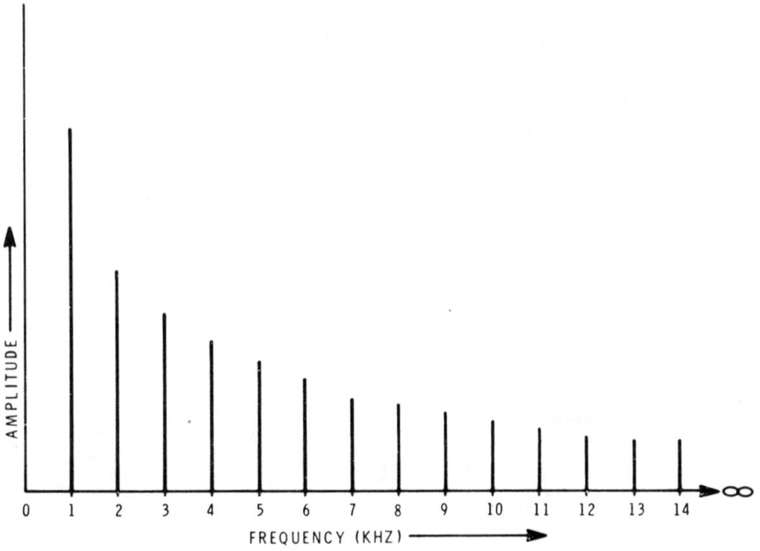

Figure 1-6
The frequency spectrum of a sawtooth wave.

AMPLITUDE MODULATION

As stated earlier, with amplitude modulation, the carrier's amplitude is varied in accordance with the modulating signal. There are several circuits that accomplish this and they will be examined in the next unit. Right now, we wish to limit our discussion to the characteristics of the modulated waveform itself.

The AM Waveform

Figure 1-7 shows a very simple AM circuit. Here, a radio frequency carrier is applied at "A" and the modulating audio tone at "B". The circuit consists of a nonlinear device such as a diode or transistor. The two signals "mix" in this circuit and produce the AM waveform shown at "C". Notice that both the negative and positive peaks of the output waveform correspond exactly to the modulating tone's waveform.

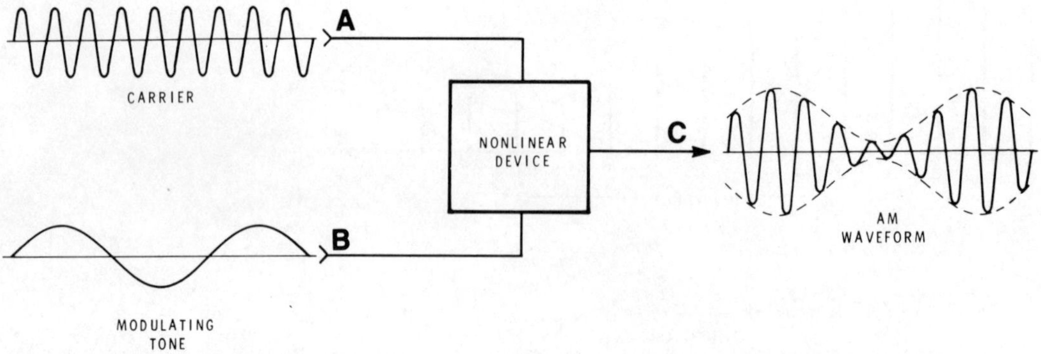

Figure 1-7
The basic method of obtaining amplitude modulation.

The amplitude and frequency of the modulating tone determines the shape of the output waveform or the **modulation envelope**. For example, Figure 1-8A shows a high amplitude audio signal. The resultant modulated waveform is shown in Figure 1-8B. On the other hand, Figure 1-8C shows a low amplitude, higher frequency audio signal. The modulated waveform is in Figure 1-8D.

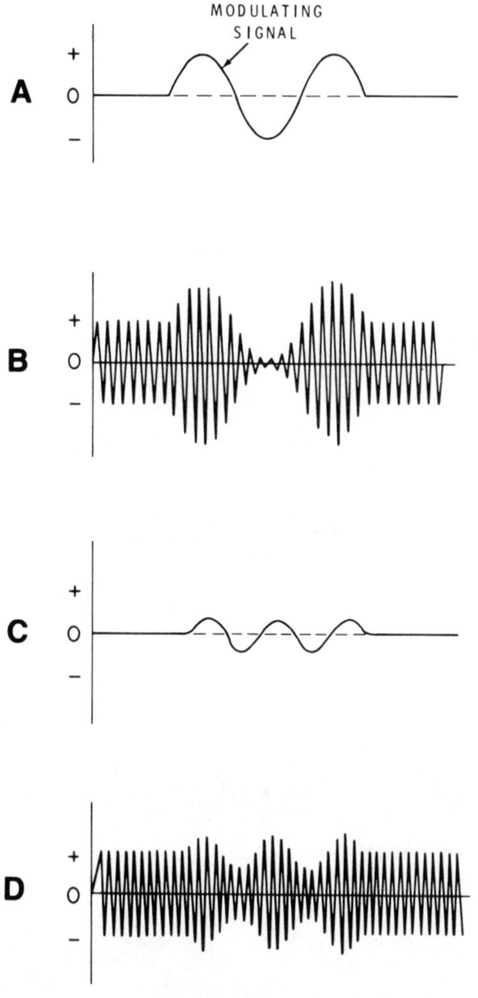

Figure 1-8
Examples of how the modulated waveform varies with the modulating signal.

Percent of Modulation

The waveforms of Figures 1-8 B and D are said to have different degrees of modulation. The degree of modulation is normally expressed as a percentage from 0 to 100. However, it is also known as the modulation factor which varies from 0 to 1. An unmodulated carrier like that shown in Figure 1-9A has 0% modulation. For comparison purposes, let's assume that the carrier has a peak-to-peak amplitude of 40 volts as shown.

Figure 1-9B shows the same carrier modulated to 100%. Here, the amplitude of the modulated waveform falls to zero volts for an instant during each cycle of the modulating wave. Also, the amplitude increases to 80 volt peak-to-peak once during each cycle of the modulating wave. The average peak-to-peak amplitude is still 40 volts.

In Figure 1-9C the carrier is shown modulated to 50%. The peak-to-peak amplitude varies from 60 volts to 20 volts. However, the average peak-to-peak amplitude is still 40 volts.

The equation for determining the percent of modulation is

$$\text{Percent of modulation} = \frac{E_{max} - E_{min}}{E_{max} + E_{min}} \times 100$$

For example, in Figure 1-9C,

$$\% = \frac{E_{max} - E_{min}}{E_{max} + E_{min}} \times 100$$

$$\% = \frac{60\text{ V} - 20\text{ V}}{60\text{ V} + 20\text{ V}} \times 100$$

$$\% = \frac{40\text{ V}}{80\text{ V}} \times 100$$

$$\% = 0.5 \times 100 = 50$$

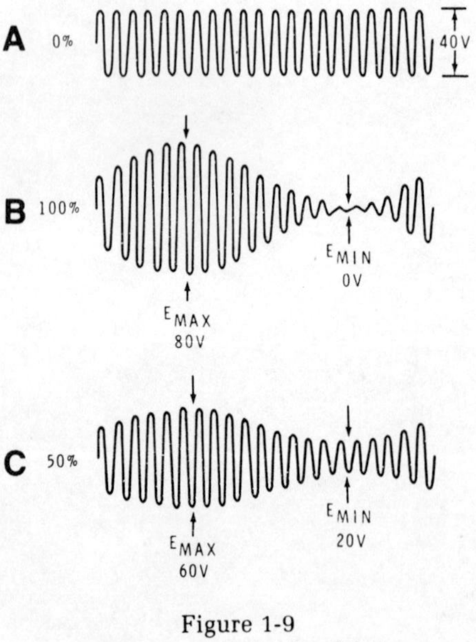

Figure 1-9
Measuring the percent of modulation.

Generally, it is desirable to keep the percent of modulation high. For a given transmitter power, a high percent of modulation will produce a stronger audio tone in the receiver. The reason for this can be visualized from Figure 1-10. Since the AM receiver recovers just the modulation envelope of the transmitted wave, it is easy to see that the higher modulated waveform in "B" will produce a louder signal than that in "A".

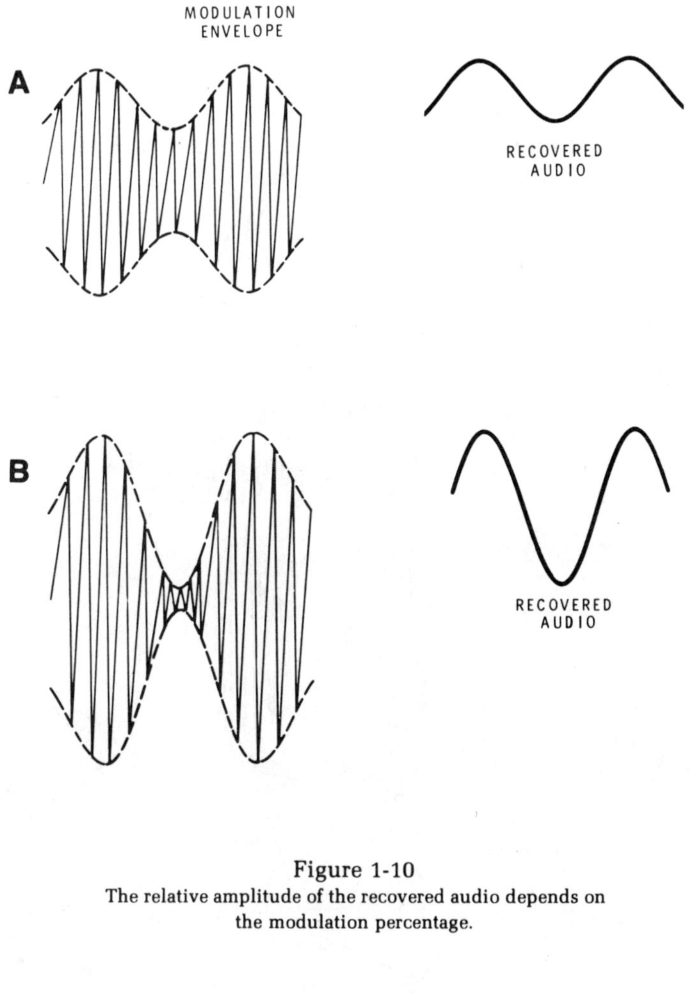

Figure 1-10
The relative amplitude of the recovered audio depends on the modulation percentage.

While it is a good idea to keep the percent of modulation high, **overmodulation** must be avoided. Overmodulation is shown in Figure 1-11C. It occurs when the amplitude of the modulating signal is too high compared to the unmodulated carrier. Obviously, the minimum amplitude of the carrier is zero volts. It cannot drop below this level regardless of how high the modulating signal is. If the modulating signal is too high, it will cause the carrier to cut off for a portion of each cycle. As a result, part of the envelope will be distorted. That is, the envelope will not be an accurate representation of the modulating wave.

Figure 1-11A shows the high modulating waveform. The unmodulated carrier is shown in Figure 1-11B. The modulated waveform, shown in Figure 1-11C, cuts off for a portion of each cycle. At the receiver, the envelope is detected, and since it is distorted, the detected waveform is also distorted. The detected envelope is shown in Figure 1-11D.

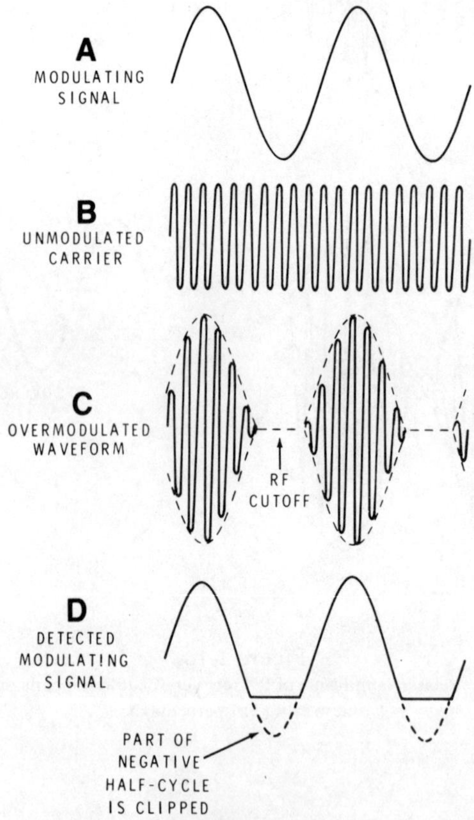

Figure 1-11
Overmodulation causes severe distortion in the received signal.

Sidebands

In the first section of this unit, you learned that any complex waveform can be broken down into its component sine waves. The same is true for an amplitude modulated waveform such as that shown in Figure 1-12A. This wave is a 1 MHz carrier modulated by a 10 kHz sine wave. At first glance, you might say that the wave is composed of a 1 MHz sine wave and a 10 kHz sine wave. However, if we apply the waveform to both a 10 kHz bandpass filter and a 1 MHz bandpass filter as shown in Figure 1-12B, we would see an output only at the 1 MHz filter. This shows that there is no 10 kHz signal present in the modulated wave, the modulation envelope only represents the audio signal. But we do know that some other sine wave components must exist in this complex wave.

If we had a tunable bandpass filter, or a spectrum analyzer, we could search the spectrum and determine what other frequencies are contained within the modulated signal. Doing this, we would find a signal at 1.01 MHz and another signal at 0.99 MHz. These two signals are called **sidebands**.

They can be extracted from the modulated waveform by using sharply tuned filters as shown in Figure 1-12C.

The higher frequency (1.01 MHz) is called the upper sideband. Its frequency is always equal to the carrier frequency plus the modulating frequency. That is:

Upper sideband $= f_c + f_m$

Where:

f_c = carrier frequency
f_m = modulating frequency

In our example:

Upper sideband = 1 MHz + 10 kHz
= 1.01 MHz

The lower frequency (0.99 MHz) is called the lower sideband. Its frequency is equal to the carrier frequency minus the modulating frequency. In this case, it is:

Lower sideband $= f_c - f_m$
= 1 MHz − 10 kHz
= 0.99 MHz

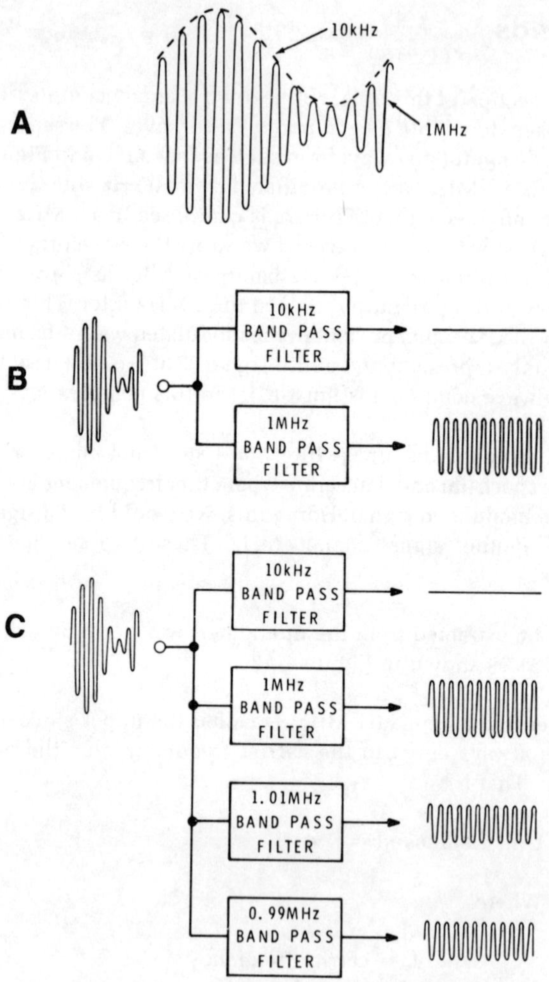

Figure 1-12
Frequency domain analysis of an AM wave.

If we were to carefully observe the filter outputs of Figure 1-12C, we would find that the sideband and carrier amplitudes do not vary. In fact, we would find that the carrier amplitude never varies whether it is modulated or not. You might ask, if the individual frequency components do not change, how can the modulated waveform change to follow the modulating signal? Figure 1-13 gives a detailed look at exactly what happens. It shows that the constant amplitude sidebands are at different frequencies and, therefore, are in phase and out of phase with one another at various times. For example, at point A, they are exactly in phase with each other. At this time, the constant amplitude carrier is also in phase. The result is a high amplitude peak in the modulated waveform. Now observe the wave relationships at point B. Once again, the sidebands are in phase with each other. However, they are 180° out of phase with the carrier wave. The result is a low point, or trough, in the modulated waveform.

From this analysis, you can see that the shape of the modulation envelope is dependent on the sidebands. And, the sidebands are in turn dependent on the modulating signal. That is, the frequency of the sidebands determines their phase relationship and, therefore, the peaks and troughs, or frequency, of the modulation envelope. The sideband amplitude will also determine the envelope's amplitude, or percent of modulation. This is because they will be either adding to or subtracting from the constant amplitude carrier. This illustrates an important fact about amplitude modulation: The modulating intelligence or information is contained only in the sidebands.

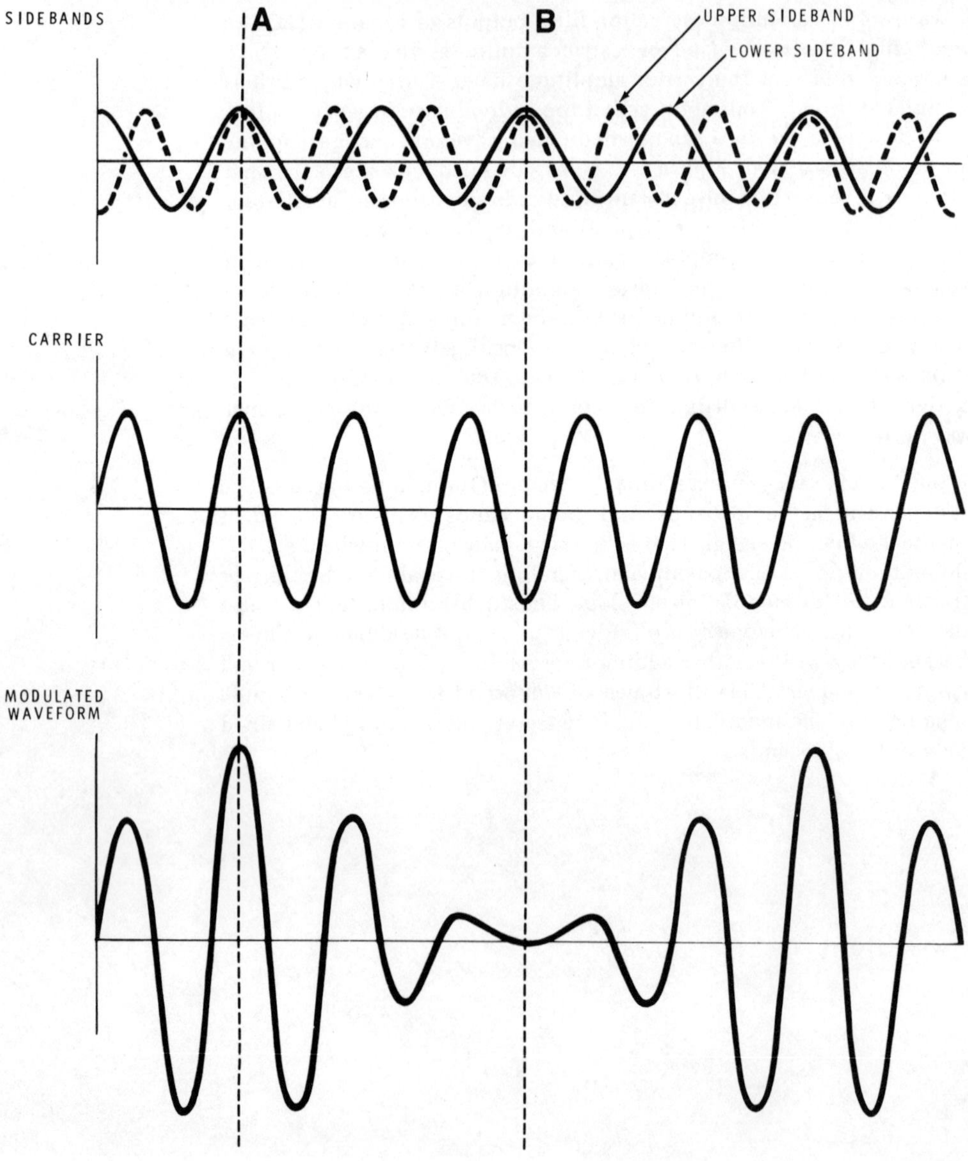

Figure 1-13
The phase relationships of an AM wave.

The frequency spectrum charts of Figure 1-14 will further illustrate this point. Since these are voltage diagrams, the sideband amplitudes shown will add or subtract directly from the carrier to produce the modulated envelope. For example, Figure 1-14A shows the sideband amplitudes as being exactly one half that of the carrier. This is the condition for 100% modulation, because when all signals are in phase, the waveform amplitude will be twice the carrier and when the sidebands are out of phase with the carrier, the waveform amplitude will be zero.

Figure 1-14B shows a 50% modulated signal. Note that the carrier amplitude remains the same while the sideband amplitudes have decreased. The frequency of the sidebands has also changed. Since the sidebands are further from the carrier, the modulating frequency has increased. This is shown in the modulated waveform to the right.

The result of a square wave modulating signal is shown in Figure 1-14C. In this case, since the modulating signal is actually the fundamental and all odd-order harmonics, there is a sideband for each sine wave in the modulating signal.

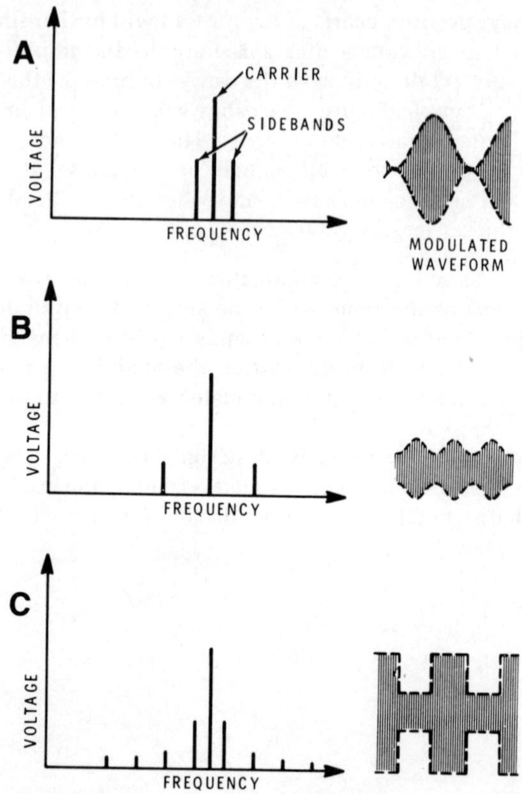

Figure 1-14
The sideband spectrum of AM waves.

Bandwidth

It is readily apparent from the frequency spectrum charts shown in Figure 1-14 that, with amplitude modulation, the transmitted signal is actually a band of frequencies rather than just the carrier. The carrier contains no information. If we transmitted or received just the carrier, no information would be conveyed. In AM systems, both the carrier and the sidebands must be transmitted and received.

The bandwidth of an AM signal extends from the lowest sideband frequency to the highest sideband frequency. Therefore, the bandwidth is always twice the highest modulating frequency. Thus, if the highest modulating frequency is 15 kHz, then the bandwidth will be 30 kHz. In the case of a complex modulating wave, such as a square wave, the bandwidth is twice the highest harmonic contained in the wave. However, each AM transmitter has bandwidth limitations above which it cannot go. In this case, the transmitter itself would limit the maximum bandwidth.

ANGLE MODULATION

There are three characteristics of a sine wave that can be modulated: amplitude, angular velocity or frequency, and phase angle. The last two are similar in that the angle, velocity or phase, is being modulated. They are also similar in that the modulation of one indirectly causes the other to vary. Therefore, they are grouped together under the heading of Angle Modulation.

Frequency Modulation

Figure 1-15 shows a frequency modulated or FM waveform. The information or modulating waveform is shown in Figure 1-15A, while the unmodulated carrier is shown in Figure 1-15B. With FM, the modulating signal changes the frequency of the carrier rather than its amplitude. The resulting frequency modulated waveform is shown in Figure 1-16C.

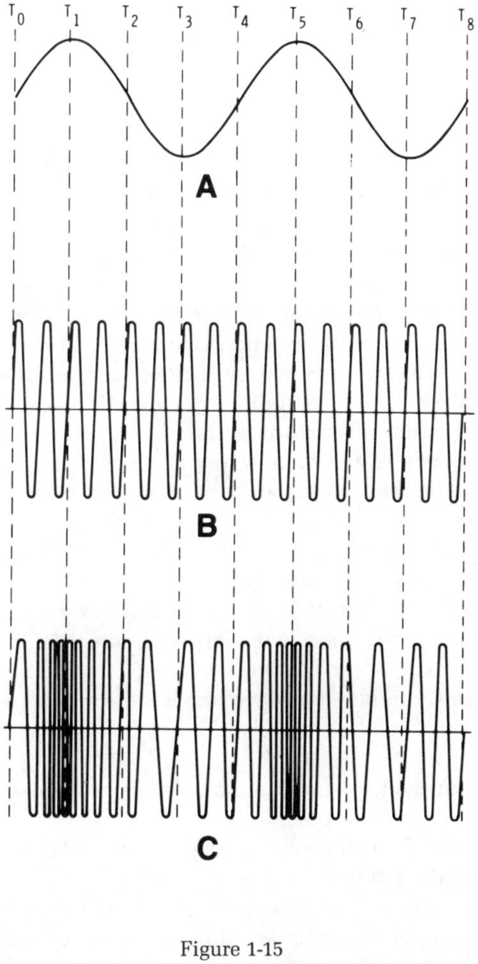

Figure 1-15
The modulating signal (A), the unmodulated carrier (B), and the frequency modulated waveform (C).

At time T_0, the modulated waveform is at its center frequency. As the modulating signal swings positive, the frequency of the carrier is increased. The carrier reaches its maximum frequency when the modulating signal reaches its maximum amplitude, at time T_1.

At time T_2, the modulating signal returns to 0 and the carrier returns to its center frequency. After T_2, the modulating signal swings negative. This forces the carrier below its center frequency. The carrier again returns to its center frequency when the modulating signal returns to 0 volts at time T_4. Between times T_4 and T_8, the modulating signal repeats its cycle. As a result, the carrier again shifted in frequency. It swings first above then below its center frequency. Notice that it returns to its center frequency each time the modulating signal passes through 0 volts.

The carrier changes equally above and below its center frequency. The amount of frequency change is called the **frequency deviation**. For example, let's assume that a carrier continuously swings from 100 MHz, down to 99.9 MHz, back to 100 MHz, up to 100.1 MHz, and back to 100 MHz. The frequency deviation is ±0.1 MHz or ± 100 kHz.

The rate of frequency deviation is determined by the frequency of the modulating signal. For example, if the modulating signal is a 1 kHz audio tone, the carrier will swing above and below its center frequency 1000 times each second. A 10 kHz audio tone will still cause the carrier to deviate ±100 kHz; but this time at the rate of 10,000 times each second. Thus, the **frequency** of the modulating signal determines the **rate** of frequency deviation but not the amount of deviation.

The **amount** that the carrier deviates from its center frequency is determined by the **amplitude** of the modulating signal. A high amplitude audio tone may cause a deviation of ±100 kHz. A lower amplitude tone of the same frequency may cause a deviation of only ±50 kHz.

Thus, the frequency-modulated waveform has the following characteristics:

1. It is constant in amplitude but varies in frequency.

2. The rate of carrier deviation is the same as the frequency of the modulating signal.

3. The amount of carrier deviation is directly proportional to the amplitude of the modulating signal.

Phase Modulation

In phase modulation (PM), the carrier's instantaneous phase angle is made to vary in proportion to the modulating signal's amplitude. However, an instantaneous phase change cannot occur without a corresponding frequency change. Therefore, phase modulation also produces carrier frequency deviation. In fact, frequency modulation cannot occur without a certain amount of phase variation. Thus, any variation in frequency or phase produces both FM and PM; the two must occur simultaneously.

A comparison of phase and frequency modulation is shown in Figure 1-16. Careful analysis of the two modulated waveforms shows them to be essentially identical. The difference is that, with phase modulation, the carrier frequency deviation is proportional to the modulation signal's phase change as well as its amplitude. For this reason, in phase modulation, maximum deviation occurs when the modulating signal crosses the zero axis. In other words, the greatest phase change occurs when the modulating wave changes polarity from + to − and from − to +.

This is shown in Figure 1-16 when, at time T_2, the modulating wave crosses the zero axis. The result is that the PM wave is deviated to its lowest frequency. At time T_4, when the modulating signal changes from − to +, the PM wave is deviated to its maximum frequency. Thus, the modulating signal, while phase modulating the carrier, also indirectly frequency modulates it. For this reason, phase modulation is sometimes referred to as **indirect FM**.

As stated earlier, with phase modulation the amount of carrier deviation is proportional to the modulating signal's phase change. This is why maximum deviation occurs when the modulating wave crosses the zero axis. However, consider what happens when the modulating frequency increases. For example, what happens if the modulating frequency increases from 100 Hz to 1000 Hz? At 100 Hz, the signal's phase changes 100 Hz × 360° or 36,000 degrees per second. At 1000 Hz, the phase change will increase by a factor of 10 to 360,000 degrees per second. Since the carrier's deviation is proportional to phase change, its deviation will increase as the modulating frequency increases.

Thus, the primary difference between PM and FM is:

1. With PM, the carrier frequency deviation is proportional to both the modulation **frequency** and **amplitude.**

2. With FM, the frequency deviation is proportional to the modulation **amplitude** regardless of the signal's frequency.

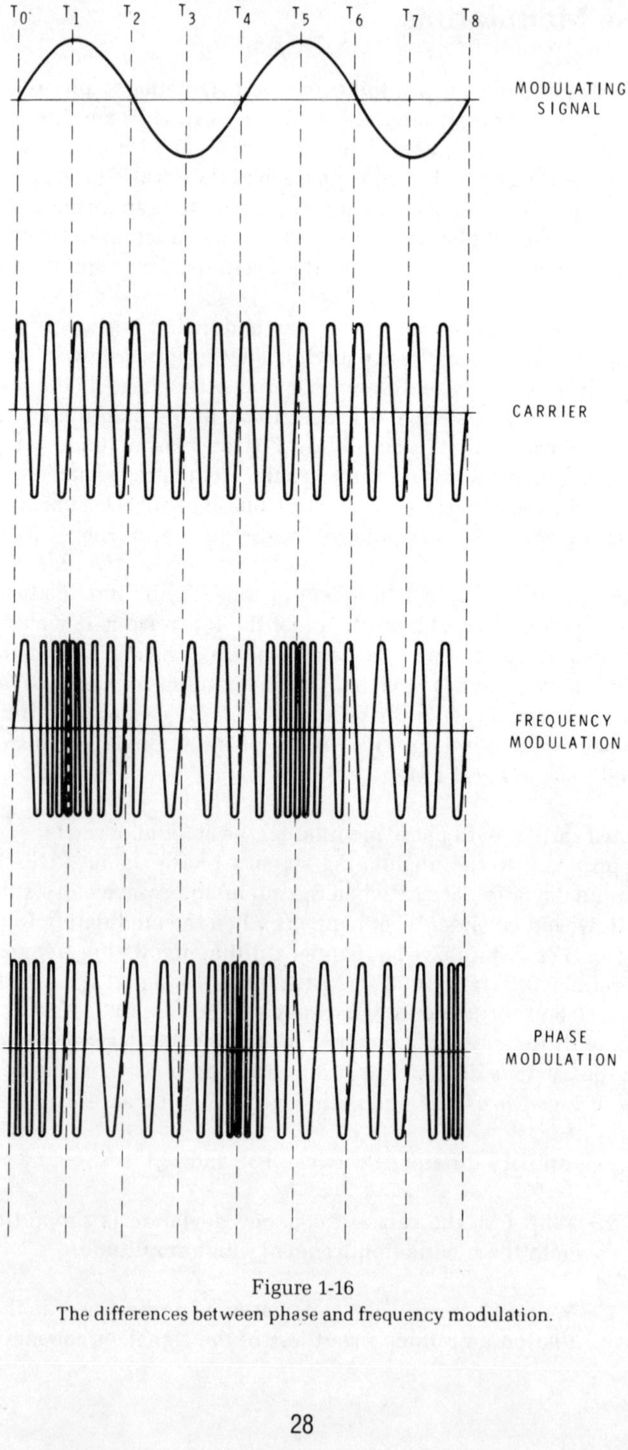

Figure 1-16
The differences between phase and frequency modulation.

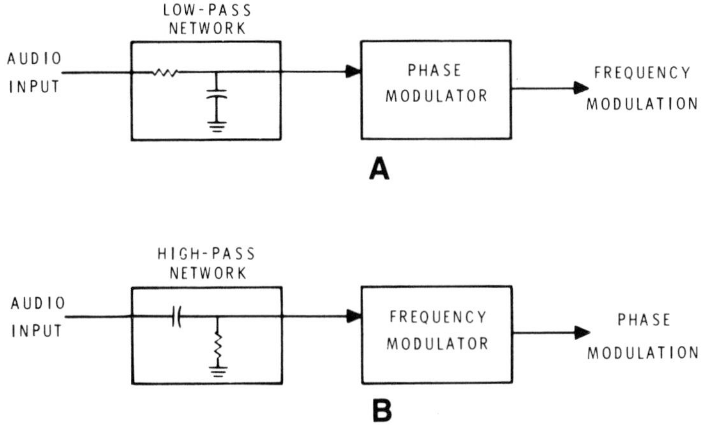

Figure 1-17
Conversion between PM and FM.

Actually, at a single modulating frequency with the same deviation, FM and PM are indistinguishable. The only difference is that as the modulating frequency increases, the PM waveform's deviation increases. With this in mind, we can design a simple RC network to convert from PM to FM or vice versa.

Figure 1-17A shows this concept in practice. Here, a low pass network attenuates the modulating signal as its frequency increases. The resulting output is then applied to a phase modulator. (The actual modulator circuits are discussed in a later unit). The output is an FM wave. This is because, as the modulating frequency increases, the low pass network correspondingly reduces its amplitude. The end result is that, as the deviation increases, due to an increase in the modulating frequency, it is also decreased due to the lower modulating signal amplitude. The two changes offset one another. Therefore, the output deviation remains constant regardless of modulating frequency. Thus, the output is actually an FM signal.

The reverse is also possible. Figure 1-17B shows a high pass network used to obtain phase modulation from a frequency modulator. The high pass filter attenuates the low frequencies and allows the high frequencies to pass. Therefore, the output wave's deviation will increase as the modulating frequency increases. The output, thus, has the characteristics of phase modulation.

It can be concluded from the foregoing that the actual differences between FM and PM are moot. Therefore, we shall disregard them for the rest of this discussion.

Modulation Index

In AM, the degree of modulation is measured as a percentage from 0% to 100% or as a modulation factor from 0 to 1. In angle modulation, the degree of modulation is measured by the **modulation index**. The equation for modulation index is:

$$m = \frac{f_d}{f_m}$$

Where

f_d = The frequency deviation

f_m = The modulating frequency

While the modulation factor in AM is limited to a decimal between 0 and 1, it must be emphasized that the modulation index in angle modulation can reach quite high numerical values. For example, the maximum deviation in FM broadcasting is 75 kHz. If a 1 kHz audio signal causes full deviation, the modulation index is:

$$m = \frac{75 \text{ kHz}}{1 \text{ kHz}}$$

$$= 75$$

Another measure of angle modulation is the **deviation ratio**. This is the ratio of the maximum deviation to the maximum audio frequency, thus, it is a total system measurement rather than the instantaneous measurement of modulation index. Using the FM broadcast system as an example, maximum deviation is 75 kHz and the maximum audio frequency is 15 kHz. The deviation ratio is:

$$\text{deviation ratio} = \frac{f_{d\ max}}{f_{m\ max}}$$

$$= \frac{75 \text{ kHz}}{15 \text{ kHz}}$$

$$= 5$$

Sidebands

One of the distinct differences between amplitude and angle modulation is the number of sidebands. Of course, AM has only two, the upper and lower sidebands. In angle modulation, the number of sidebands is theoretically infinite. This is because the numerous frequency shifts produced by the modulating signal cause the generation of many additional frequencies. Fortunately, many higher order sidebands contain an insignificant amount of energy and can, therefore, be disregarded.

In AM, the sidebands add to or subtract from the constant amplitude carrier which results in the modulation envelope. However, in angle modulation, the waveform remains at a constant amplitude regardless. This means that, as sideband number, amplitude, or distribution change, the carrier must also change to keep the resultant waveform's amplitude constant. This interelationship between carrier and sidebands is orchestrated by the modulation index. That is, the modulation index determines the number of significant sidebands, their amplitude, and the carrier's amplitude.

A graph of modulation index versus carrier and sideband amplitude is shown in Figure 1-18. You can see that, with a modulation index of zero, the carrier amplitude is one and there are no sidebands. As the modulation index increases, more sidebands are added, their amplitude increases, and the carrier decreases. This happens until, at an index of 2.4, the carrier disappears entirely. This is known as the first carrier "null", since it also occurs at still higher modulation indices. At an index of 3.1 the carrier's amplitude is −0.3. This indicates that it is 180° out of phase with the components above the zero axis.

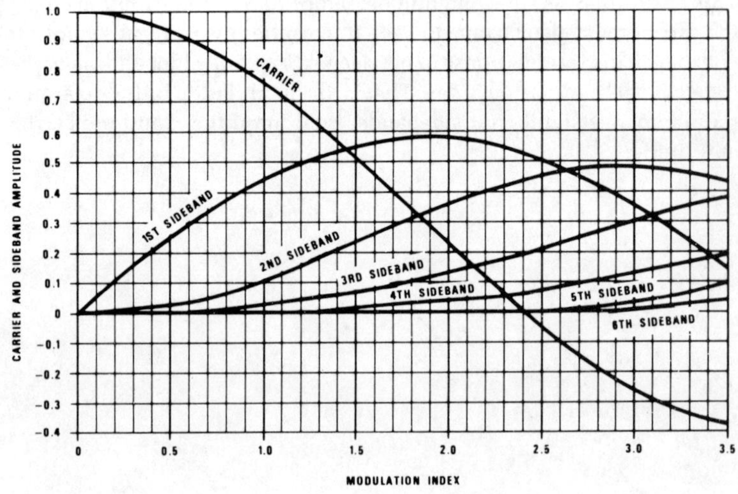

Figure 1-18
Graph of sideband and carrier amplitude in an angle modulated signal.

MODULATION INDEX	CARRIER	SIDEBANDS															
		1st PAIR	2nd	3rd	4th	5th	6th	7th	8th	9th	10th	11th	12th	13th	14th	15th	16th
0.00	1.00	—	—	—	—	—	—	—	—	—	—	—	—	—	—	—	—
0.25	0.98	0.12	—	—	—	—	—	—	—	—	—	—	—	—	—	—	—
0.5	0.94	0.24	0.03	—	—	—	—	—	—	—	—	—	—	—	—	—	—
1.0	0.77	0.44	0.11	0.02	—	—	—	—	—	—	—	—	—	—	—	—	—
1.5	0.51	0.56	0.23	0.06	0.01	—	—	—	—	—	—	—	—	—	—	—	—
2.0	0.22	0.58	0.35	0.13	0.03	—	—	—	—	—	—	—	—	—	—	—	—
2.5	-0.05	0.50	0.45	0.22	0.07	0.02	—	—	—	—	—	—	—	—	—	—	—
3.0	-0.26	0.34	0.49	0.31	0.13	0.04	0.01	—	—	—	—	—	—	—	—	—	—
4.0	-0.40	-0.07	0.36	0.43	0.28	0.13	0.05	0.02	—	—	—	—	—	—	—	—	—
5.0	-0.18	-0.33	0.05	0.36	0.39	0.26	0.13	0.05	0.02	—	—	—	—	—	—	—	—
6.0	0.15	-0.28	-0.24	0.11	0.36	0.36	0.25	0.13	0.06	0.02	—	—	—	—	—	—	—
7.0	0.30	0.00	-0.30	-0.17	0.16	0.35	0.34	0.23	0.13	0.06	0.02	—	—	—	—	—	—
8.0	0.17	0.23	-0.11	-0.29	-0.10	0.19	0.34	0.32	0.22	0.13	0.06	0.03	—	—	—	—	—
9.0	-0.09	0.24	0.14	-0.18	-0.27	-0.06	0.20	0.33	0.30	0.21	0.12	0.06	0.03	0.01	—	—	—
10.0	-0.25	0.04	0.25	0.06	-0.22	-0.23	-0.01	0.22	0.31	0.29	0.20	0.12	0.06	0.03	0.01	—	—
12.0	-0.05	-0.22	-0.08	0.20	0.18	-0.07	-0.24	-0.17	0.05	0.23	0.30	0.27	0.20	0.12	0.07	0.03	0.01
15.0	-0.01	0.21	0.04	0.19	-0.12	0.13	0.21	0.03	-0.17	-0.22	-0.09	0.10	0.24	0.28	0.25	0.18	0.12

Figure 1-19
Table of sideband and carrier distribution for several modulation indices.

Figure 1-19 is a table of sideband and carrier amplitudes at various modulation indices. Although the sidebands theoretically stretch out to infinity, any sidebands with an amplitude less than 1% of the original carrier are insignificant and, therefore, left out. As an example, suppose the modulation index is 0.5. From the table in Figure 1-19, a modulation index of 0.5 means that carrier amplitude is 0.94, the first sidebands have an amplitude of 0.24, and the second, and last significant sidebands, have an amplitude of 0.03. This sideband distribution is shown in Figure 1-20 along with several others.

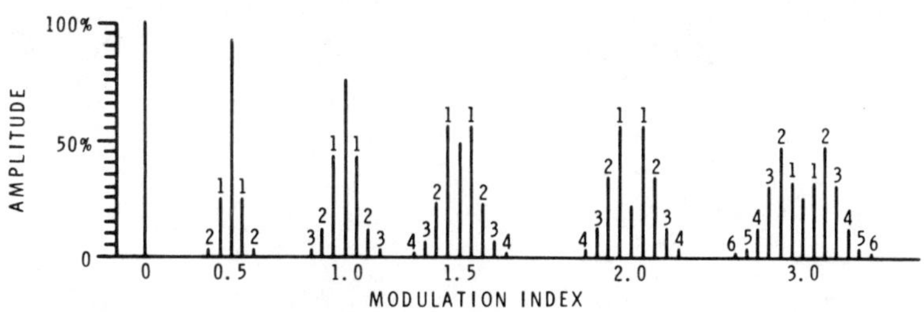

Figure 1-20
Sideband distribution for angle modulation.

Bandwidth

In angle modulation, the number of sidebands and their amplitude is determined by the modulation index. However, the frequency of each sideband depends on the modulating frequency. The first-order sidebands are $f_c + f_m$ and $f_c - f_m$. The second-order sidebands are $f_c + 2f_m$ and $f_c - 2f_m$. This progression continues for each higher order sideband. The bandwidth, therefore, depends on the number of sidebands in the wave. This is determined by the modulating frequency and modulation index. Since modulation index is $\frac{f_d}{f_m}$, frequency deviation is also a determining factor in bandwidth.

Thus, if we know the modulating frequency and the frequency deviation, we can easily determine the required bandwidth.

As an example, what is the bandwidth of an angle modulated signal in which the modulating frequency is 3 kHz and the maximum deviation is 18 kHz? We must first find the modulation index.

$$m = \frac{f_d}{f_m}$$

$$= \frac{18 \text{ kHz}}{3 \text{ kHz}}$$

$$= 6$$

From the table in Figure 1-19, we find that a modulation index of 6, has 9 significant sideband pairs. Therefore, the bandwidth is

$$BW = f_m \times \text{highest order sideband} \times 2$$

$$= 3 \text{ kHz} \times 9 \times 2$$

$$= 54 \text{ kHz}$$

THE COMMUNICATIONS SYSTEM

So far in this unit, we have discussed the need for modulation and the characteristics of the two basic types of modulation. This section is a brief overview of the communications system itself.

The basic block diagram of a communications system is shown in Figure 1-21. It consists of an input device, a transmitter, a transmission channel — which also introduces noise, a receiver, and an output device. Let's discuss each of these items.

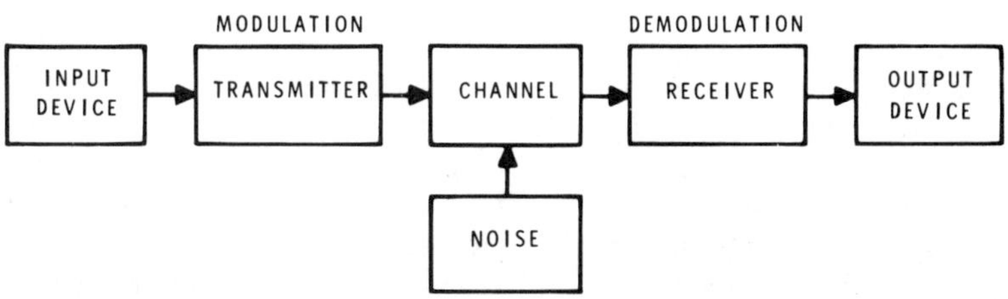

Figure 1-21
A communications system.

Input/Output Devices

An input or output device is nothing more than a **transducer**, which is the name given to any device that converts an input of one physical form to an output of a different physical form. For example, in radio broadcasting the input device is a microphone and the output device is a loudspeaker. Both of these are transducers. The microphone converts sound waves into electrical signals, while the loudspeaker converts the electrical signals into sound waves.

The same is true for television broadcasting. The visual input device is a TV camera which converts light into electrical signals. The output device is the cathode ray, "picture," tube which converts the electrical signals into light.

There are a number of other input/output devices used in communications. These include telegraph keys, teletypewriters, chart recorders, facsimile machines, and, in telemetry systems, a wide variety of transducers. Many of these items require a complete course of study on their own. However, they all have an electrical output or input which attaches to the transmitters and receivers you will study in this course.

Transmitter

As stated earlier, it is usually impractical to transmit the input signal at its original frequency. Therefore, the transmitter must transfer the input signal to a higher frequency. It does this through the process of modulation. Thus, the transmitter consists of a high frequency carrier generator and a modulator. It may also use amplifiers to increase its power output.

Channel

The transmission channel can be any radio frequency between approximately 3 kHz and 300 GHz. It could also be any light wave frequency. However, this course deals only with radio frequency communications.

Figure 1-22 shows the radio frequency spectrum. It is divided into several frequency bands from VLF (very low frequencies, 3 kHz-30 kHz) to EHF (extremely high frequencies, 30 GHz-300 GHz). Each of these bands has specific characteristics. For example, the medium frequency (MF) band allows local communications during daylight hours and longer range communications at night. The standard AM broadcast band is located within this band. The international short-wave broadcasting band is located in the HF band (3 MHz-30 MHz). These frequencies allow world-wide communications and, therefore, are the best for international broadcasting. The television channels are located in the VHF and UHF bands. These frequencies offer local fade-free communications and, therefore, are well suited to TV broadcasting.

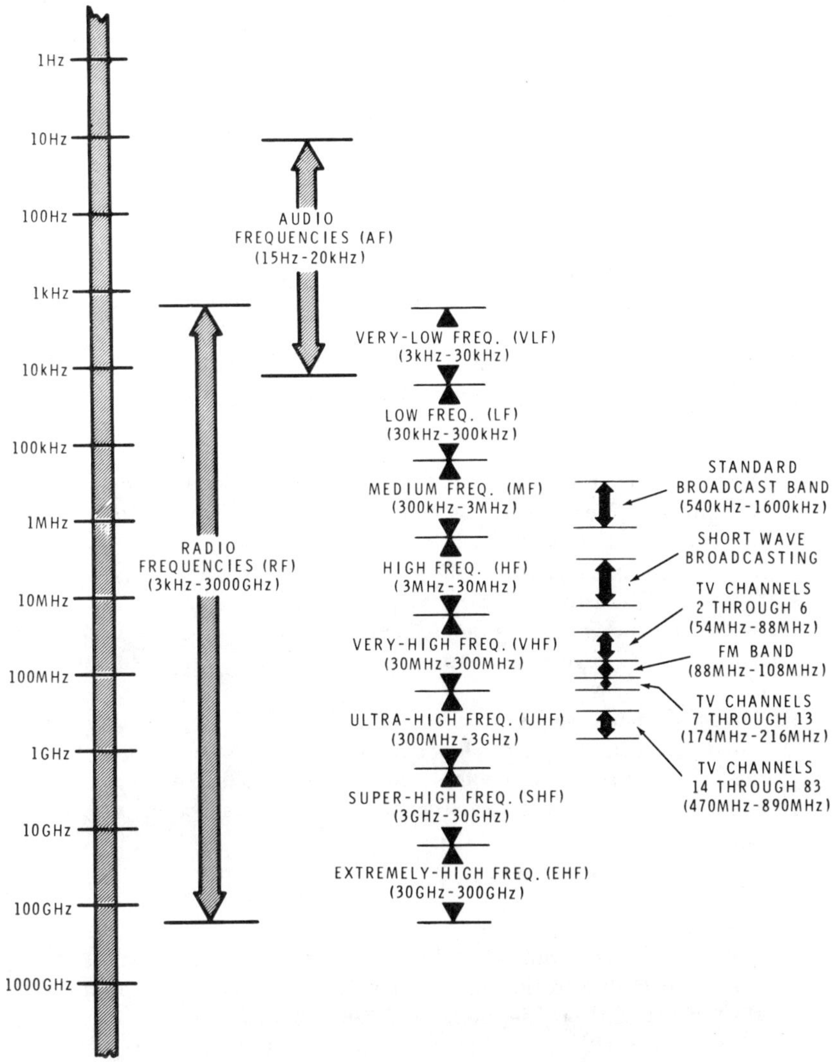

Figure 1-22
The radio frequency spectrum.

Noise

Noise is defined as any unwanted form of energy tending to interfere with the reception of wanted signals. In Figure 1-24, only one source of noise is shown, not because only one exists, but to simplify the drawing. There are basically four types of noise. These are atmospheric noise, space noise, man-made noise, and internal noise.

Atmospheric noise is caused by electrical disturbances in the earth's atmosphere. The most visible of these sources is lightning. Since radio waves can propagate around the world, given the right conditions, so can the noise generated by lightning. Therefore, atmospheric noise due to lightning is present at all times, even though there may be no local thunderstorm in progress.

Space noise is due primarily to the sun and the billions of other stars in the universe. The sun is an extremely large body at a very high temperature of over 6000°C on the surface. It therefore radiates energy over a very broad frequency spectrum. This is a constant source of noise. However, there are also disturbances on the sun's surface such as sun spots and flares that generate even higher levels of noise. Since distant stars are also suns, they too have high temperatures and, therefore, radiate noise.

Almost all of us are familiar with man-made noise. The electric motor of a hair dryer, vacuum, etc., can literally wipe out reception. Other sources of man-made noise are automobile ignitions, high voltage power line leakage, fluorescent lights, and other transmissions can be considered as noise.

Internal noise is created by the active and passive devices inside the receiver itself. This type of noise can be divided into two categories: thermal noise and shot noise. Thermal noise is generated by any resistance and is due to the rapid and random motion of the atoms and electrons of which the resistance is constructed. Shot noise is present in all active devices and is due to the random variation in the arrival of electrons at the collector of a transistor or the drain of an FET.

Receiver

By the time the transmitted signal reaches the receiver, most of its power has been dissipated and several forms of noise have been added. The receiver must amplify this weak signal enough to overcome any internal noise and also separate it from all the other transmitted signals present. It must then demodulate this signal to recover the original information. This is quite a difficult task. Over the years, many types of receivers have been developed to perform these functions for specific applications. We will discuss many of these receivers in a later unit.

Unit 2

AMPLITUDE MODULATION

INTRODUCTION

In the previous unit, you studied the need for modulation, its various types, and its characteristics. This unit discusses several methods of generating amplitude modulation. It also discusses the disadvantages of AM and presents a more advanced modulation technique — single sideband.

AM CIRCUITS

To obtain amplitude modulation, the audio signal and the carrier must be combined in a stage called the **modulator**. This circuit is used in basically two types of devices — signal generators and AM transmitters. A signal generator produces a low-level AM signal for testing and alignment purposes. This type of modulator is relatively simple. However, an AM transmitter must produce power levels from 5 watts, for a citizen's band (CB) transceiver, up to 2 million watts, for a broadcast transmitter. Such high power levels require very complex modulators, primarily to improve efficiency. This section discusses both types of modulators.

The Diode Modulator

A simple circuit for producing amplitude modulation is shown in Figure 2-1. The modulating signal, in this case audio, is applied at the top of R_1 while the carrier is applied at the top of R_2. The signal at the junction of R_1 and R_2 is the sum of the carrier and audio. That is, the carrier is simply riding on the audio signal. Notice that the carrier is not amplitude modulated at this point. It is simply added to the audio signal.

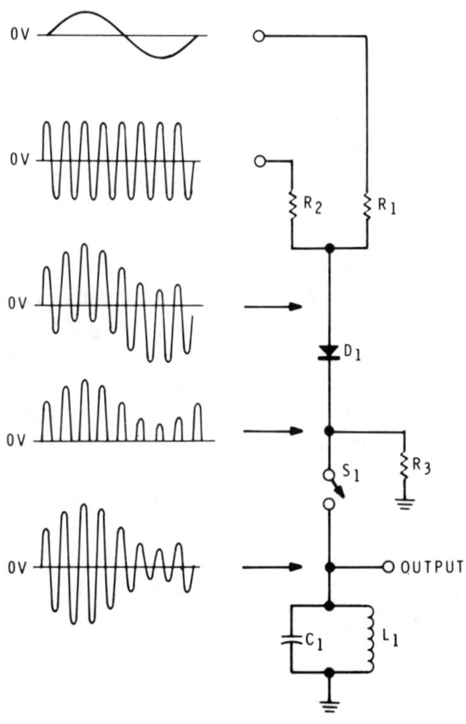

Figure 2-1
The diode modulator and its waveforms.

When S_1 is open, D_1 conducts through R_3 when the signal on the anode swings positive. However, when this signal swings negative, D_1 cuts off. Thus, the signal developed across R_3 will consist of positive-going pulses as shown. Notice that the positive pulses vary in amplitude in accordance with the audio signal.

S_1 is included in the circuit merely for explanation purposes. Normally, it is omitted and the tank circuit composed of C_1 and L_1 is connected directly in parallel with R_3.

The purpose of the tank circuit, which is tuned to the carrier frequency, becomes clear when S_1 is closed. Each time D_1 conducts, a pulse of current flows through the tank. This causes the tank to resonate and the flywheel action of the tank produces a negative half-cycle for each positive input pulse. The high amplitude positive pulses cause high amplitude negative pulses, and the low amplitude positive pulses cause low amplitude negative pulses. Therefore, each negative half-cycle will have the same amplitude as the positive half-cycle. As you can see, the output is an AM wave. Thus, this simple circuit produces amplitude modulation.

The diode modulator is a good example of **nonlinear mixing**. That is, the audio and carrier are mixed across the nonlinear resistance of D_1. The result of this mixing action is sum and difference products. Therefore, the signals across D_1 are the original frequencies and their sum and difference. If the carrier is 100 kHz and the audio signal is 10 kHz, the sum is 100 kHz + 10 kHz = 110 kHz, and the difference is 100 kHz − 10 kHz = 90 kHz. These are the sideband frequencies.

The 100 kHz carrier and 10 kHz audio signal are also present. However, the resonant circuit of C_1-L_1 is tuned to 100 kHz and, therefore, allows only the carrier and sidebands to pass to the output. The 10 kHz audio signal is filtered out. Thus, the output of the diode modulator is a true amplitude modulated signal.

The Absorption Modulator

The absorption modulator is used primarily in VHF, UHF, and higher frequency signal generators. Most absorption modulators use PIN diodes because of their variable impedance characteristics. Usually, several of these diodes are arranged in an attenuator network. This arrangement not only allows amplitude modulation, but also a variable output level control which makes it an extremely versatile device for use in signal generators.

Another advantage of the absorption modulator is that it does not affect the frequency generating device or circuit. This is very important at VHF, UHF, and higher frequencies since any voltage changes at the oscillator may cause undesired frequency variations.

You will recall that a PIN diode is formed by placing an intrinsic (I) semiconductor material between the P and N type materials of a junction diode. Above 100 mHz, this causes the diode to act like a linear resistance by conducting in both directions. However, its resistance depends on the amount of forward bias current. Therefore, the PIN diode acts as a current-controlled resistance. This makes it ideal for attenuator networks.

Figure 2-2 shows a typical attenuator. The variable resistor elements represent PIN diodes. A DC bias current is applied to the diodes, which varies their resistance and, thus, the network's attenuation of the RF signal. Therefore, the steady-state DC bias current can be used to adjust the output amplitude and function as the signal generator's output level control. Similarly, the modulation signal can be superimposed on the DC bias and, as a result, amplitude modulate the RF signal.

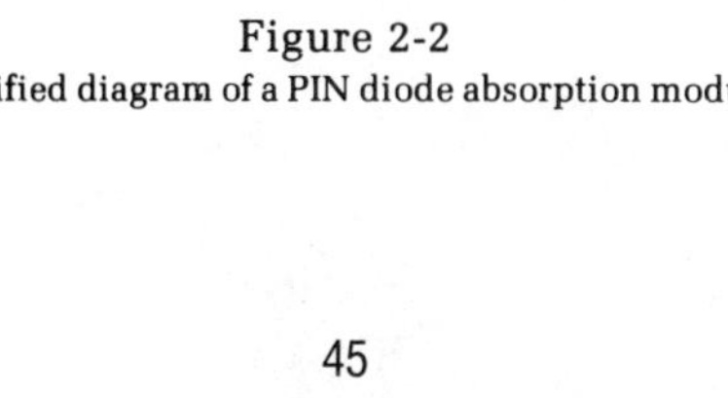

Figure 2-2
A simplified diagram of a PIN diode absorption modulator.

However, for 100% amplitude modulation the positive peak of the waveform must be twice the carrier level. For this reason, the attenuator must be set for approximately mid-range attenuation. This establishes the carrier level and allows equal amplitude variations both above and below this set level.

The efficiency of the absorption modulator is very low. However, at microwave frequencies it is one of the few methods of amplitude modulation available. Moreover, since it is normally used at low power levels, efficiency is not an important requirement.

The Collector Modulator

The most commonly used AM technique is collector modulation or, in vacuum tube modulators, plate modulation. In collector modulation, the modulating signal is applied to the transistor's collector, in series with the DC supply voltage.

A typical collector modulation circuit is shown in Figure 2-3. Q_1, and its associated circuitry, is a class C RF power amplifier. Note that the modulating signal is applied to the collector through transformer T_1. With no modulating signal present, there is approximately zero volts dropped across the secondary of T_1. Therefore, the entire V_{cc} voltage is present at the collector of Q_1. However, when modulation is applied, there is an AC voltage, E_m, induced in the secondary of T_1. Since it is an AC voltage, it is alternately in phase and out of phase with V_{cc}. When it is in phase with V_{cc}, a high positive potential is applied to the collector of Q_1. This causes an increase in collector current and, therefore, a higher amplitude output signal. This is the peak of the modulation envelope.

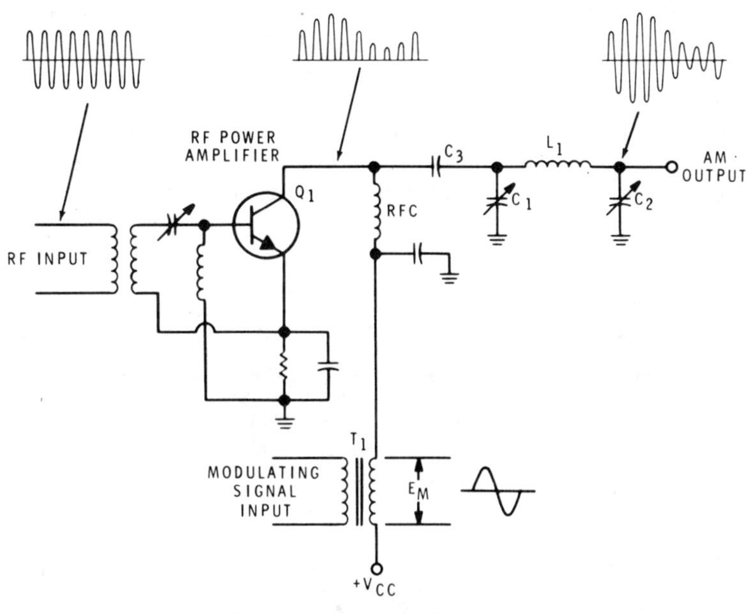

Figure 2-3
Basic collector modulation circuit.

When E_m is out of phase with V_{cc}, a lower positive potential is applied to Q_1. This results in lower collector current and, hence, a lower output amplitude. This is the trough of the modulation envelope.

Since Q_1 is a class C amplifier, only positive pulses will appear at its output. To form a complete AM waveform, these output pulses must be applied to a resonant or "Tank" circuit. The circuit of L_1, C_1, and C_2 in Figure 2-3 form the output tank for this circuit. You may recognize this circuit as a pi network. A similar circuit is often used as a low pass filter in power supplies. The circuit shown here is also a low pass filter, but it is different from the power supply filter in two important ways: it is a resonant tank circuit, and an impedance matching network. Let's examine its operation in detail.

Figure 2-4A shows the basic output network. The output signal of Q_1 is developed across the RF choke (RFC) and capacitively coupled to the pi network by C_3. If the schematic is rearranged, as shown in Figure 2-4B, you can see that the combination of C_1 and C_2 are in parallel with L_1. Therefore, by tuning C_1 and C_2, the circuit will resonate at the output frequency. Figure 2-4B also shows that the input is across C_1 while the output is taken across C_2. By varying the ratio between C_1 and C_2, an impedance match can be obtained between Q_1 and the output load. Thus, not only is the pi network a low-pass filter, which reduces harmonic output, but it is also a resonant tank and impedance matching network.

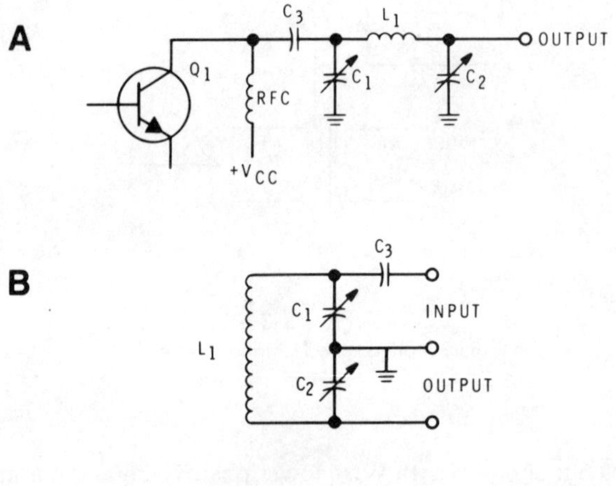

Figure 2-4
Pi output tank circuit.

One problem with the collector modulation circuit of Figure 2-3 is that 100% modulation is not possible because the transistor saturates before reaching the positive peak of modulation. Also, at the negative trough, the base RF drive feeds through the base-collector junction capacitance, thereby preventing zero output at the negative peak. Both of these problems can be overcome by modulating both the driver and final amplifier. This is shown in Figure 2-5. By modulating the driver, extra drive is provided to the final amplifier at the positive peak, allowing it to reach a full 100% modulation. At the negative trough, the drive is reduced sufficiently to prevent RF feedthrough.

Notice that the modulator in the circuit of Figure 2-5 is a push-pull audio amplifier. It must deliver at least one-half of the input carrier power to obtain 100% amplitude modulation. Therefore, a relatively high power audio amplifier is required. The push-pull circuit delivers high power at much better efficiency than a single-ended class A amplifier. Let's examine exactly what the modulator power requirements are.

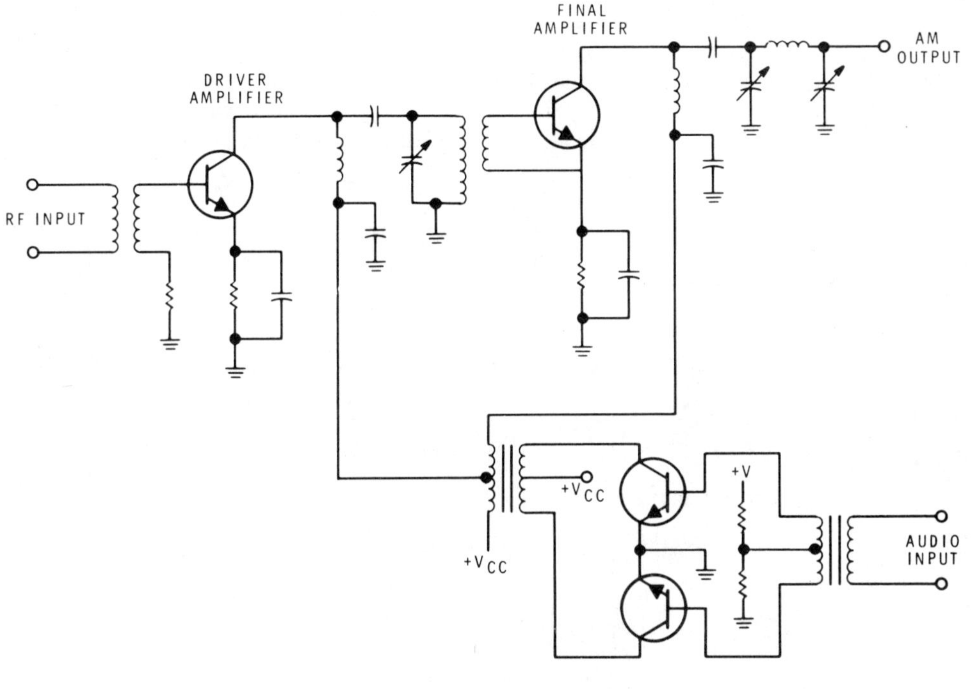

Figure 2-5
Collector modulation of both driver and final amplifiers.

As an example, suppose that V_{cc} is 28 volts and I_c is 1 ampere. The DC input power is:

$$
\begin{aligned}
P_{IN} &= V_{cc} \times I_c \\
&= 28\ V \times 1\ A \\
&= 28\ W
\end{aligned}
$$

The impedance of the collector circuit is:

$$Z_c = \frac{V_{cc}}{I_c}$$

$$= \frac{28\ V}{1\ A}$$

$$= 28\ \Omega$$

This is the impedance that the audio modulating power must be developed across.

To modulate the amplifier stage 100%, the sum of the instantaneous audio peak voltage and V_{cc} must be twice the V_{cc} for the positive peak and zero for the negative trough. Since V_{cc} is 28 V, the modulator must double this by supplying 28 V **peak** to the transformer secondary. This voltage must be developed across the 28 Ω collector impedance. To find the power required, you can use the formula $P = \frac{E^2}{R}$. However, this equation applies only to effective or rms values of AC voltages. Therefore, the peak audio voltage must be converted to rms:

$$E_{rms} = E_{PK} \times 0.707$$

$$E_{rms} = 28\ V \times 0.707$$

$$= 19.8\ V$$

Now, using the power formula:

$$P = \frac{E^2}{R}$$

$$= \frac{19.8\ V^2}{28\ \Omega}$$

$$= 14\ W$$

Thus, the modulator output must be 14 W to obtain 100% modulation of a 28 W RF amplifier.

From these figures you can also determine the sideband and carrier power in the transmitted signal. Continuing our example, if the efficiency of the final amplifier is 75%, the carrier output power is:

$$P_{out} = P_{in} \times EFF\%$$

$$= 28\ W \times 75\%$$

$$= 28\ W \times 0.75$$

$$= 21\ W$$

At 100% modulation, the modulator supplies 14 W to the final amplifier. The amount of modulator power that appears at the output is also determined by the final amplifier efficiency. Therefore,

$$P_{out} = P_{in} \times EFF\%$$

$$= 14\ W \times 0.75$$

$$= 10.5\ W$$

The modulator power supplied to the output, which is divided equally between the sidebands, is 10.5 W or one-half the carrier output power.

An equation for determining modulator power or sideband power for any modulation percentage is:

$$P_m = \frac{m^2 P_c}{2}$$

Where: P_m = modulator power or sideband power

m = percent of modulation

P_c = carrier power.

To determine modulator power, P_c must be the carrier **input** power. To find sideband power, P_c must be the carrier **output** power.

Let's try a couple of examples. What is the sideband power for a 100-watt carrier at 50% modulation?

$$P_m = \frac{m^2 P_c}{2}$$

$$P_m = \frac{0.5^2 \times 100 \text{ W}}{2}$$

$$= 12.5 \text{ W}$$

Since sideband power is 12.5 W, the power of each sideband is 6.25 W.

What modulator power is required for 80% modulation of an RF amplifier with an input power of 1000 W?

$$P_m = \frac{m^2 P_c}{2}$$

$$= \frac{0.8^2 \times 1000 \text{ W}}{2}$$

$$= 320 \text{ W}$$

Series Modulator

While collector modulation is the most commonly used technique, it does have disadvantages. The most serious of these is the modulation transformer itself. It is large, heavy, costly, and introduces frequency distortion. That is, due to the inductance of the transformer, some frequencies are attenuated more than others. One method that eliminates the modulation transformer is series modulation. This technique uses a transistor in place of the transformer.

A basic series modulator is shown in Figure 2-6. In this circuit, the current through the final amplifier, Q_1, is controlled by the series modulator, Q_2. Therefore, any increase in current flow through Q_2 will increase the output of Q_1. Likewise, any decrease in Q_2 current flow will decrease the output of Q_1. Thus, the series modulator amplitude modulates the RF signal present at Q_1 and it does it through direct coupling rather than transformer coupling. This eliminates any frequency distortion.

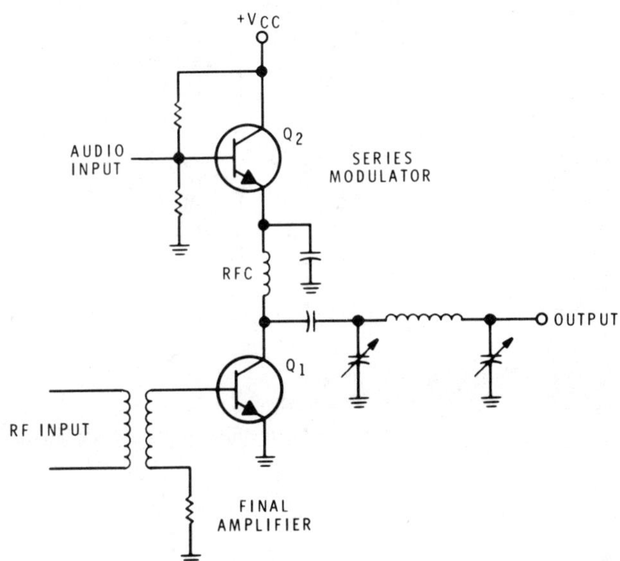

Figure 2-6
A basic series modulation circuit.

However, series modulation has one significant disadvantage: it is very inefficient. You will recall that at the positive peak of 100% modulation, the voltage at the final amplifier must be doubled. As an example, if the final amplifier has a collector voltage of 100 V and collector current of 1 A, the input power is 100 W. In order to 100% modulate this amplifier, the collector voltage and current must double for the positive peak of modulation. This additional voltage must be dropped across the series modulator at 0% modulation. Therefore, at 0% modulation, the series modulator drops 100 V at 1 A. This is shown in Figure 2-7. Note that the modulator is dissipating a full 100 W. This is very inefficient operation. Collector modulation is much more efficient, since at 0% modulation, there is almost no voltage dropped across the secondary of the modulation transformer.

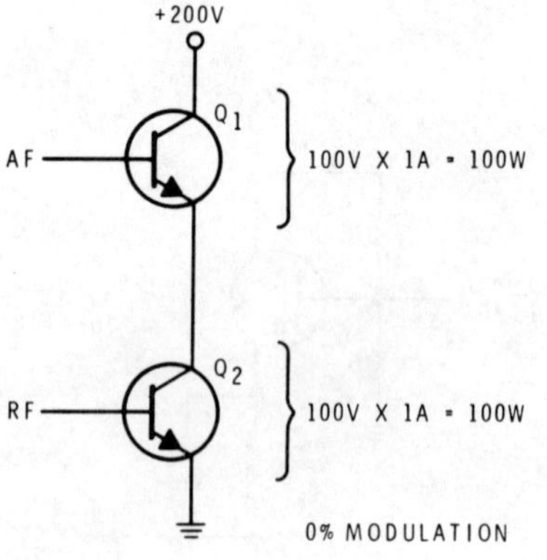

Figure 2-7
Power distribution at 0% modulation in a series modulator.

Progressive Series Modulator

One circuit that greatly improves the efficiency of series modulation while still maintaining the advantage of direct coupling is the progressive series modulator. This circuit is shown in Figure 2-8. It uses two modulator transistors and two power supplies in a special switching configuration. At 0% modulation and during the negative troughs of modulation, Q_1 is turned off and, therefore, the 104 V supply is disconnected from the circuit. During the negative troughs, Q_2 controls the current and voltage applied to the final amplifier. For positive modulation, Q_2 saturates and Q_1 turns on. As soon as Q_1 turns on, D_1 is reverse biased and, therefore, disconnects the 52 V supply. At this time, Q_1 controls the voltage and current to the final amplifier. The 104 V supply is now connected, which allows a full positive peak of modulation.

The advantage of this arrangement is that, at carrier-only conditions or 0% modulation, very little power is dissipated across the modulator. This is because, at 0% modulation, Q_1 is off while Q_2 is on and saturated. Its saturation voltage is approximately 2 V, therefore, this leaves 50 V from the 52 V supply for the final amplifier. Thus, at carrier-only conditions, 50 V is applied to the final amplifier. If the current is 2 A, input power is 50 V × 2 A = 100 W. However, the power dropped across modulator transistor Q_2 is only 2 V × 2 A = 4 W. This is a signficant difference over the normal series modulator. Thus, the progressive series modulator offers much better efficiency while maintaining direct coupling. The only disadvantage is the increased circuit complexity.

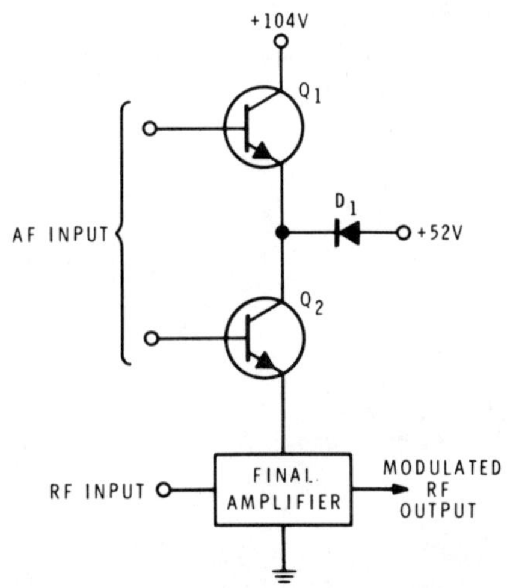

Figure 2-8
A simplified diagram of a progressive series modulator.

SUPPRESSED CARRIER AM

The type of amplitude modulation just discussed will be referred to in this section as standard AM. In this system, the carrier and both sidebands are transmitted just as they appear at the output of the modulator. This system is used by standard AM broadcast stations and most citizen's band radios. Its primary advantage is that it uses straightforward and inexpensive transmitting and receiving equipment. However, it has several disadvantages. One of these disadvantages can be overcome by transmitting a suppressed carrier signal.

This section discusses the disadvantages of AM, methods of generating suppressed carrier AM, and its disadvantages.

Disadvantages of Standard AM

The standard AM system has several disadvantages. The three most important are:

1. Most of the transmitted power is in the carrier. This power is wasted because it does not contribute to the intelligence being conveyed.

2. The bandwidth of the transmitted signal is twice that of the intelligence being conveyed.

3. The sidebands and carrier must have precise amplitude and phase relationships. These relationships are difficult to maintain under some conditions.

Let's discuss each of these in more detail.

CARRIER POWER

Let's assume that a standard AM transmitter produces a 100-watt output when unmodulated. The entire 100 watts is contained in the unmodulated carrier. If the transmitter is modulated at 100%, the total radiated power will increase to 150 watts. The transmitter still produces the 100-watt carrier. The additional 50 watts represents the power in the sidebands. Since the power of the lower sideband is equal to that of the upper sideband, each sideband contributes only 25 watts to the total transmitted power. When the percent of modulation is lower than 100%, the power in the sidebands is even lower.

It is important to remember that the carrier itself does not vary in frequency or amplitude. Consequently, the carrier does not contain any intelligence. All information being transmitted is in the sidebands. The carrier acts only as a reference frequency. It allows a modulator to convert low frequency audio to high frequency sidebands. It also allows the demodulator (or detector) to recover the audio information. Between the modulator stage in the transmitter and the detector in the receiver, the carrier serves no useful purpose. Thus, in the standard AM system, at least two-thirds of the transmitted power is taken by a signal which conveys no intelligence.

BANDWIDTH

In the previous section, we discussed how the bandwidth is determined by the intelligence. When a 600 kHz carrier is amplitude modulated by a 5 kHz audio tone, a 595 kHz sideband and a 605 kHz sideband are produced. In standard AM, both the sidebands are transmitted along with the carrier. Thus, the transmitted bandwidth is 605 kHz − 595 kHz = 10 kHz. This is twice the modulating frequency.

When the modulating frequency changes, both sidebands change frequency. When the amplitude of the modulating signal changes, the power in the sidebands change. In both cases, the carrier remains constant in both power and frequency. Obviously then, the intelligence is contained in the sidebands and not in the carrier. Just as important, the intelligence is duplicated in the sidebands. That is, if we know the frequency of either sideband and the frequency of the carrier, we can determine the modulating frequency.

Therefore, in the standard AM system, the bandwidth is twice as wide as is necessary.

PROPAGATION PROBLEMS

For perfect reception in a standard AM system, the two sidebands and the carrier must be received exactly as they are transmitted. Unfortunately, when propagation conditions are poor, the AM signal can greatly deteriorate before it reaches the receiver. Moreover, each component of the AM signal may be affected differently since they have different frequencies. Thus, the standard AM system is subject to fading and interference when propagation conditions are poor.

Balanced Modulators

The first disadvantage of standard AM, wasted carrier power, can be eliminated by suppressing or removing the carrier before transmission. This is accomplished by using a balanced modulator. This device mixes the audio and RF carrier. Then, due to its circuit configuration, it allows only the mixing products or sidebands to pass to the output. The resultant signal is called double sideband suppressed carrier. It can be transmitted and received without a carrier and, thus, offers more efficient transmission than standard AM.

There are several different types of balanced modulators. Each type has its own advantages and disadvantages. Let's discuss several of the more common balanced modulators.

TWIN FET BALANCED MODULATOR

A relatively simple balanced modulator is shown in Figure 2-9. Here, the RF carrier input is applied in phase to the gates of Q_1 and Q_2. This causes equal but opposite drain currrents to flow in the primary of T_2. The result is zero output when the carrier alone is applied to the circuit. Potentiometer R_1 is used to compensate for differences in the two FETs. The output is monitored and R_1 is adjusted for minimum or zero output with the carrier only applied to the circuit.

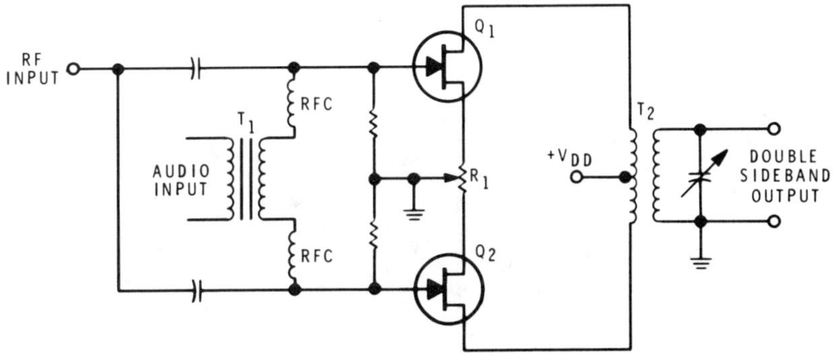

Figure 2-9
The twin-FET balanced modulator.

When audio is present, it is applied to the gate of Q_1 180° out of phase with the gate of Q_2 due to the coupling transformer T_1. This sets up a push-pull type circuit for the audio. It also causes nonlinear mixing or modulation across the FETs. The end result is that the mixing products or sidebands are in push-pull across the primary of T_2 and, therefore, pass to the output. However, the carrier is cancelled across the primary and, thus, does not appear at the output. The original audio is filtered out by the RF transformer and resonant tank. Thus, the output is a double sideband, suppressed carrier signal.

The amount of carrier suppression depends on how closely the FETs match and also the other components in the system. As mentioned before, R_1 can be adjusted to improve the balance. Also, the center tap on the primary of T_2 can be adjusted for optimum operation. When properly constructed and adjusted, this circuit can suppress the carrier to −60 dB below the sideband amplitude.

DIODE-RING BALANCED MODULATOR

A balanced modulator that uses diodes as the nonlinear elements is shown in Figure 2-10. In this circuit, the diodes must be carefully matched to achieve a good carrier balance or null.

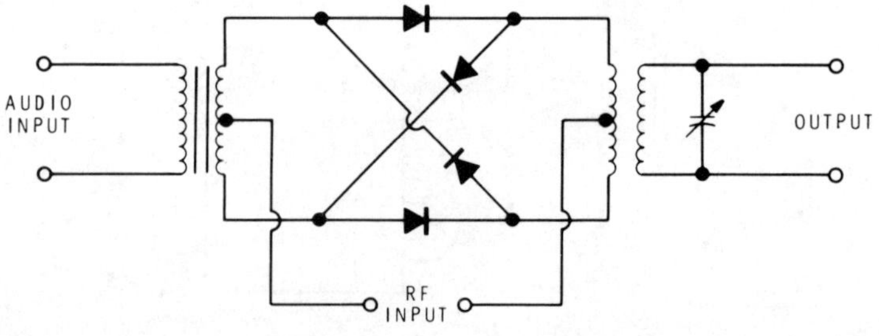

Figure 2-10
The diode-ring balanced modulator.

Figure 2-11A shows the circuit operation when the RF input has a polarity as shown. Diodes D_1 and D_2 are forward biased and equal but opposite currents flow in both sides of T_1. Therefore, there is no output. When the RF input polarity reverses, diodes D_3 and D_4 conduct as shown in Figure 2-11B. Here again, equal but opposite currents flow in T_1 and there is no output.

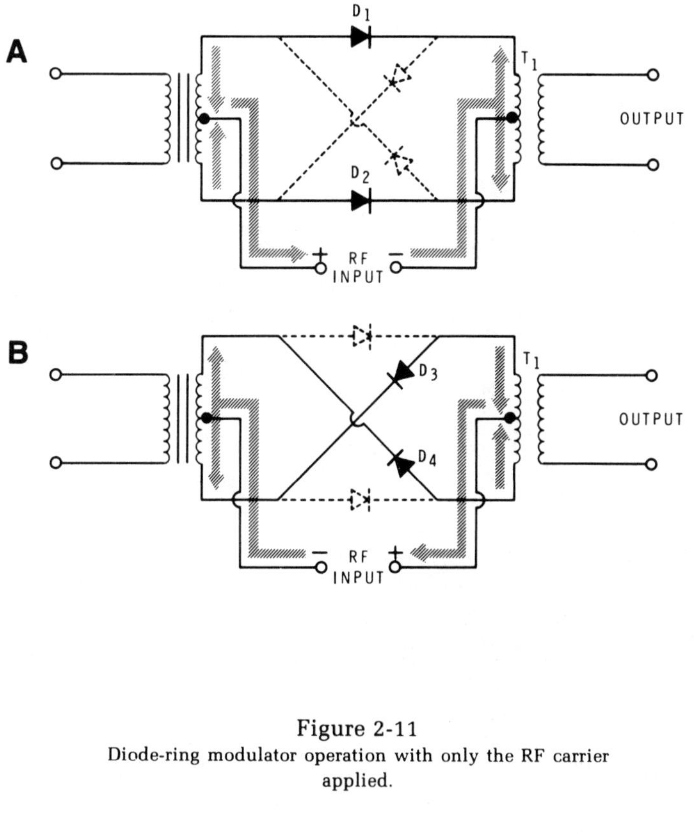

Figure 2-11
Diode-ring modulator operation with only the RF carrier applied.

Figure 2-12A shows the circuit action when the audio signal is applied. To simplify the drawing, the RF input has been removed and the diodes replaced by switches. When diodes D_1 and D_2 conduct, due to the RF input, the audio signal flows through T_1 in the direction shown. When the RF signal changes polarity, as shown in Figure 2-12B, diodes D_3 and D_4 conduct. This causes the signal through T_1 to change polarity. Thus, the audio polarity at T_1 is actually being switched back and forth by the RF input.

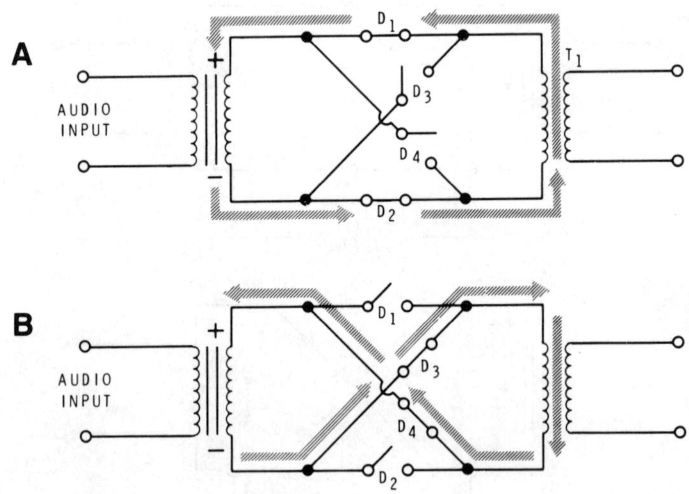

Figure 2-12
Audio signal switching action in the diode-ring modulator.

To clarify matters, the audio signal does not actually forward bias the diodes. Instead, the RF input switches the diodes and the audio signal sees the forward biased diodes as a low impedance path. This is due to the varactor action of the diodes. A forward biased diode has a high capacitance and, therefore, a low capacitive reactance to the audio signal. On the other hand, the reverse biased diodes offer a high reactance to the audio signal.

The various waveforms for this circuit are shown in Figure 2-13. The RF input is shown as a square wave. This allows instantaneous diode switching and thereby simplifies the drawing. Note that each time the RF input goes negative, the output wave reverses polarity. This is due to the diode switching action described earlier. You can follow this throughout the output waveform.

Notice that the output waveform's envelope is a "double or mirrior image" of the modulating audio. This is significantly different than the standard AM envelope which duplicates the modulating signal. It can be shown that this complex wave is actually composed of two equal amplitude sine waves whose frequency separation is twice the modulating audio frequency. This corresponds exactly to upper and lower sidebands spaced around a suppressed carrier. In short, the output is a double sideband, suppressed carrier signal.

You might well ask: where did the sidebands come from when all the circuit did was switch diodes on and off? Since the diodes are being switched, they have a nonlinear resistance. Therefore, nonlinear mixing occurs, creating the mixing products and, thus, sidebands. You will recall that the carrier is balanced out by circuit action.

The diode-ring modulator is widely used. In fact, it is often made in special packages using hot-carrier diodes. These devices offer wide operating frequency ranges and excellent carrier suppression.

There are several variations of the diode-ring balanced modulator. They all use diodes as the nonlinear elements and operate in a similar manner. That is, they balance out or null the carrier, and the nonlinear mixing across the diodes creates sidebands which are passed to the output.

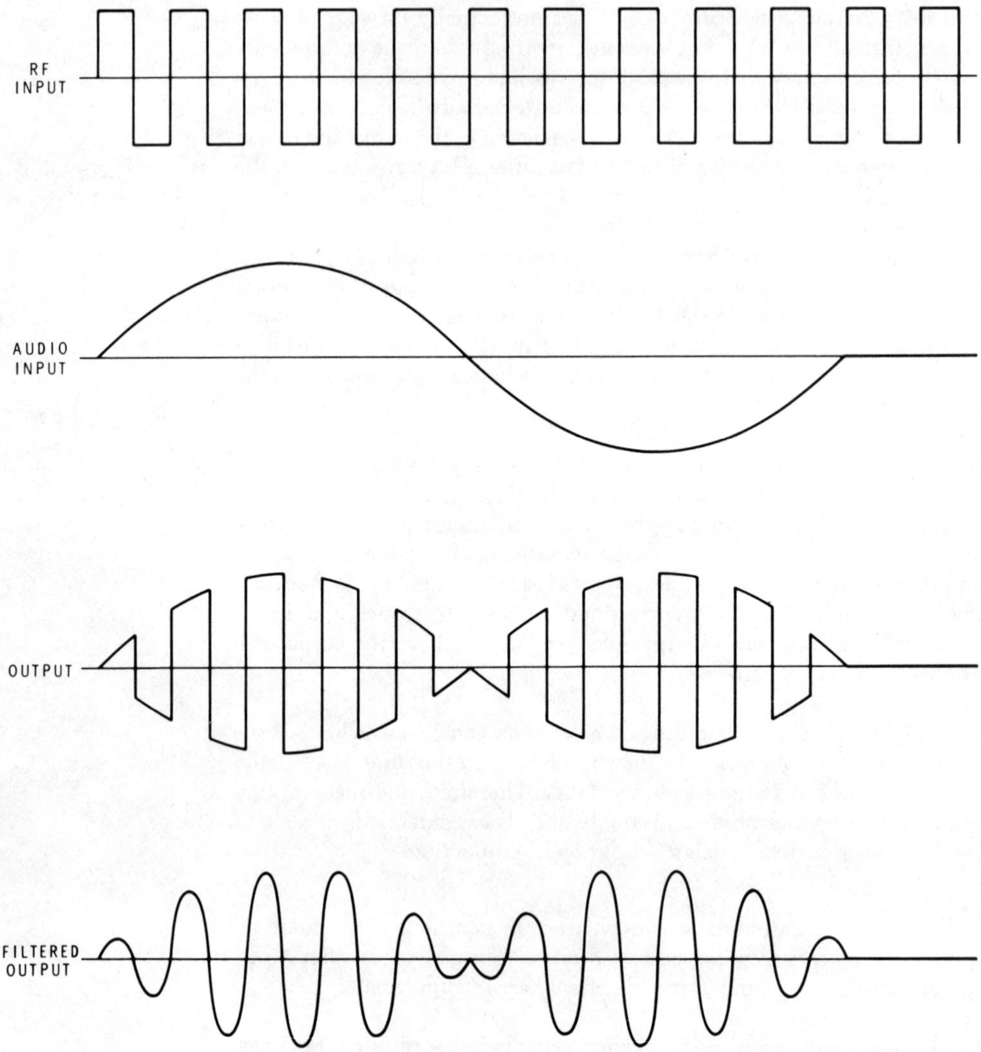

Figure 2-13
Waveforms in the diode-ring balanced modulator.

IC BALANCED MODULATORS

One circuit that lends itself extremely well to balanced modulator applications is the differential amplifier. A simplified diagram of a differential amplifier is shown in Figure 2-14. Q_3 acts as the current source for Q_1 and Q_2. If the RF input is applied to the bases of Q_1 and Q_2 in phase, current through both transistors will be identical and the voltage difference across the output will be zero. This is the common-mode rejection of the differential amplifier and it has balanced out the carrier.

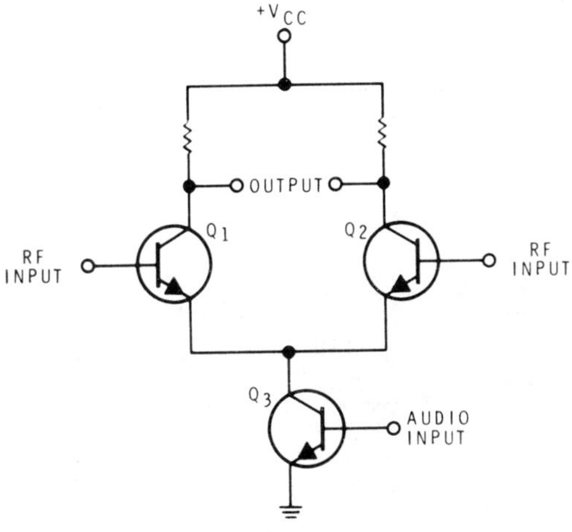

Figure 2-14
A differential amplifier used as a balanced modulator.

The audio input is applied to the base of Q_3. This upsets the circuit balance. As a result, the audio and RF signals are mixed across Q_1 and Q_2. This is nonlinear mixing and, therefore, sidebands appear at the output. However, the carrier or RF input does not. Since it is a common-mode signal, it is rejected.

A differential amplifier must be constructed of transistors whose characteristics are very closely matched. Forming the transistors on a single chip of silicon, as is done with ICs, ensures this necessary match. Therefore, the differential amplifier is ideally suited to integrated circuit construction.

Figure 2-15 shows ICs that have been specifically designed for use as balanced modulators. Figure 2-15A is the 1496 balanced modulator which is manufactured by Motorola, National, and Signetics. This device uses a differential amplifier configuration similar to what was previously described. Its carrier suppression is rated at a minimum of −50 dB with a typical value of −65 dB at 500 kHz.

Figure 2-15B is the SL1640 balanced modulator manufactured by Plessey. It also uses a differential amplifier configuration. Its primary advantage is the small number of external components required. Its carrier suppression is typically −40 dB at 30 MHz.

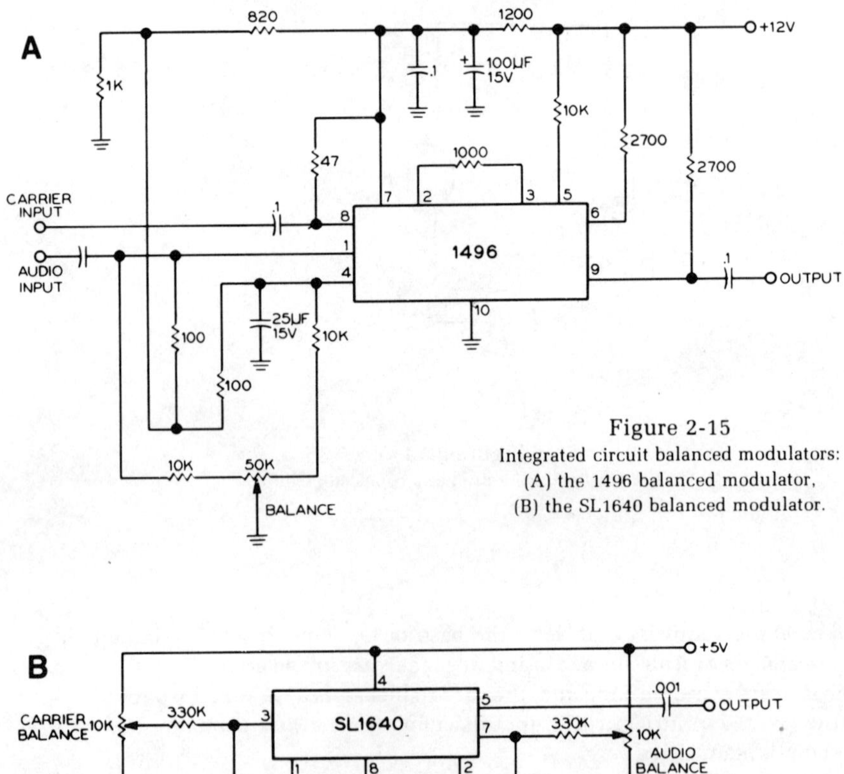

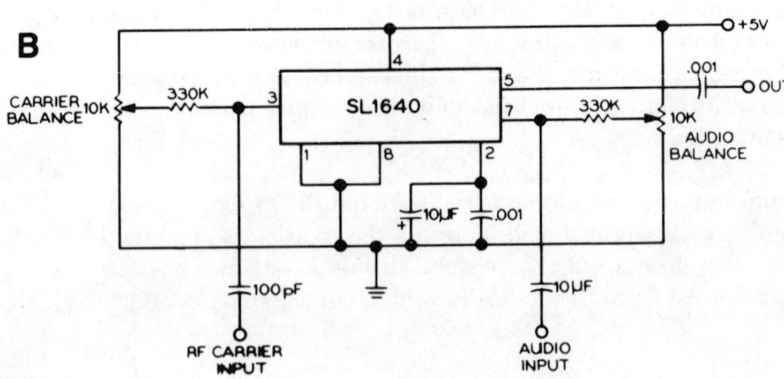

Figure 2-15
Integrated circuit balanced modulators:
(A) the 1496 balanced modulator,
(B) the SL1640 balanced modulator.

Disadvantages of Suppressed Carrier AM

One of the disadvantages of standard AM is that most of the transmitted power is in the carrier. This power is wasted since it contains no intelligence. For this reason, the balanced modulator is used to suppress the carrier. This allows transmission of only the sidebands. However, the other disadvantages of AM remain: excessive bandwidth and propagation problems.

The bandwidth of a suppressed carrier AM signal is still twice that of the highest modulating signal. This is twice as much bandwidth as is needed.

As mentioned before, for perfect AM reception, the two sidebands and carrier must be received exactly as they are transmitted. However, in suppressed carrier AM, the sidebands are transmitted without the carrier. This means that the carrier must be "reinserted" at the receiver in order to demodulate the signal. Since phase relationships are critical in AM, the reinserted carrier must not only be the same frequency as the original but also exactly in phase with the original. This is the most difficult problem to solve when designing a double sideband suppressed carrier system. It can be solved, but unfortunately, to date, the required circuitry has proven too complex and expensive for practical applications. Although, double sideband suppressed carrier signals are used as part of the color television system and the FM stereo broadcasting system.

A further disadvantage of suppressed carrier AM is that it is normally a **low-level modulation** technique. This means that it is generated at a low signal level and then must be amplified to a level suitable for transmission. Since these amplifiers must duplicate both the RF signal and the envelope, they must operate at class B or, more frequently, class AB. This results in low efficiency at high power levels. On the other hand, standard AM is usually a **high-level modulation** technique. It is normally generated at the final amplifier or last stage of the transmitter. Therefore, the RF carrier can be amplified using high efficiency class C amplifiers. However, the audio signal still requires relatively low efficiency class AB push-pull amplifiers.

As you can see, there are several trade-offs involved in the use of any system. However, the increased complexity of the double sideband suppressed carrier AM system limits its use to specialized applications.

SINGLE SIDEBAND

All three disadvantages of standard AM can be overcome by a system called single sideband suppressed carrier (SSB). In this system, only one sideband is transmitted. The carrier and the other sideband are suppressed at the transmitter. This system can still convey intelligence since either sideband contains all the information of the original modulating signal.

This approach requires a more complicated and, therefore, more expensive transmitter. The transmitter generally uses a balanced modulator to suppress the carrier and one of several methods to suppress the unwanted sideband. The design of the receiver is also complicated. As in double sideband suppressed carrier, a substitute carrier must be added to the received sideband. However, in SSB, the reinserted carrier must only be within a few Hertz of the original carrier. This is much easier to obtain than the precise frequency and phase requirements for double sideband suppressed carrier AM. As a result, SSB is less complex. Also, its many advantages over standard AM outweigh its additional cost and complexity.

The first advantage of single sideband is that all the power transmitted represents intelligence. Compare this to the standard AM system in which at least two thirds of the transmitted power is in the carrier. This means that a single sideband transmitter that is radiating 50 watts can produce the same audio signal level at the receiver as a standard AM transmitter that is radiating 150 watts.

Single sideband also reduces the bandwidth requirements. Since only one sideband is transmitted, the bandwidth is only one half that of a standard AM transmitter. Thus, a given frequency range can accommodate twice as many SSB channels as standard AM channels.

Finally, SSB is far superior to AM when propagation conditions are poor. SSB can be up to eight times more effective in getting through when propagation conditions are really bad. This is due primarily to its narrow bandwidth and the power advantages discussed above.

Filter Method SSB Transmitter

The most widely used technique for generating a single sideband signal is the filter method. A block diagram of this type of transmitter is shown in Figure 2-16. Here, the modulating audio is amplified and, along with the RF carrier, applied to the balanced modulator. The output of the balanced modulator, which is a double sideband suppressed carrier signal, is then applied to the sideband filter. This filter is designed to pass the desired sideband and block the unwanted sideband. The output is, therefore, either an upper sideband signal or a lower side band signal, depending on the filter's passband.

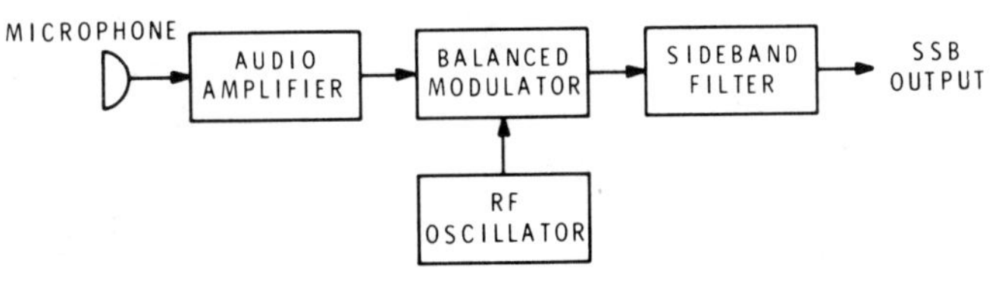

Figure 2-16
Filter method SSB transmitter.

SIDEBAND FILTERS

In a single sideband transmitter, the input audio is usually limited to the 250 Hz to 3000 Hz range. This is all that is required for adequate speech intelligibility and it narrows the signal bandwidth still further. It is accomplished by using a simple audio filter in the input amplifiers. The lower limit of 250 Hz results in a 500 Hz separation between the upper and lower sidebands as shown in Figure 2-17A.

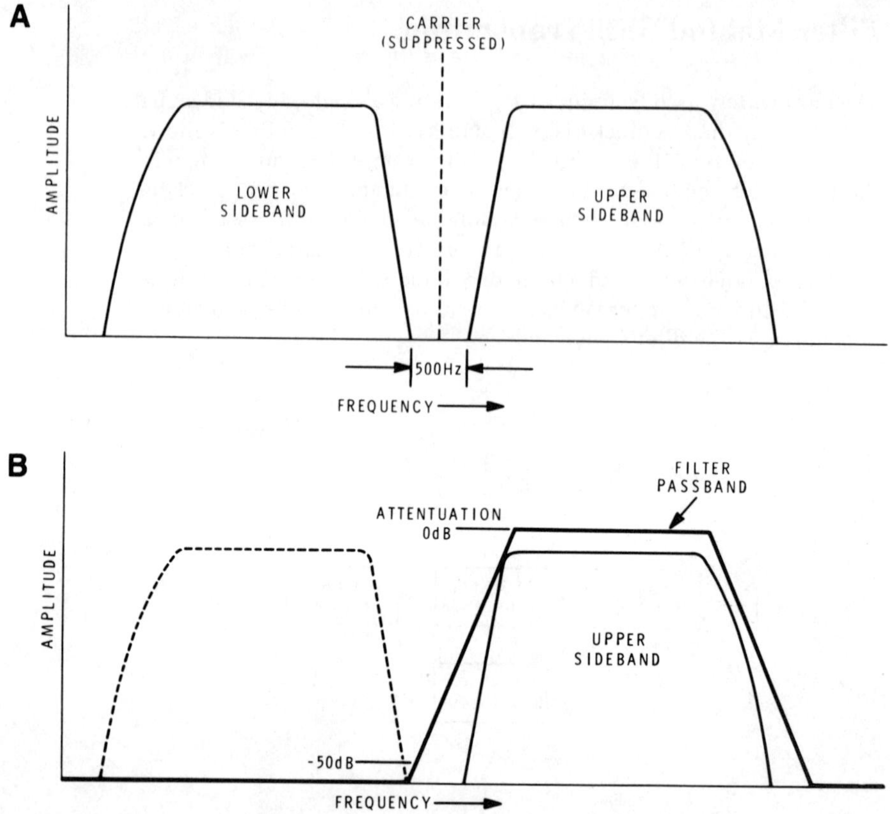

Figure 2-17
SSB filter passband.

The function of the sideband filter is to fully attenuate one sideband while allowing the other sideband to pass. Therefore, the filter must drop from 0 dB attenuation to approximately −50 dB attenuation within the 500 Hz frequency range that separates the sidebands. This requires a very sharp selectivity "skirt" or drop from 0 dB to − 50 dB. A typical sideband filter selectivity, or response, curve is shown in Figure 2-17B. In this case, the filter allows the upper sideband to pass and attenuates the lower sideband.

One of the most widely used sideband filters uses quartz crystals to achieve the desired passband shape. Let's briefly review the characteristics of crystals before continuing.

Figure 2-18A shows the equivalent electrical network for a quartz crystal. The inductance L represents the mass of the crystal, C_1 represents the piezoelectric resilience of the crystal, R is the frictional losses, and C_2 is the capacitance of the metal electrodes holding the crystal. In most crystals, the frictional losses (R) are extremely small and, therefore, the circuit Q is very high. In fact, the Q ranges from 10,000 up to 1 or 2 million. This makes them ideally suited for use in filters.

The impedance curve for a quartz crystal is shown in Figure 2-18B. This shows that at low frequencies, the network impedance is capacitive and high. As frequency increases, the impedance drops until it reaches f_S. This is the series resonant frequency of the crystal which is determined by the crystal's cut and thickness. Going back to the equivalent circuit, the reactances of L and C_1 are equal, at f_S, forming a series resonant circuit. At a slightly higher frequency, f_P, the inductive reactance of L is equal to X_{c2}. Therefore, at this frequency, the crystal is a parallel resonant circuit. These characteristics of the quartz crystal can be used to form a sideband filter.

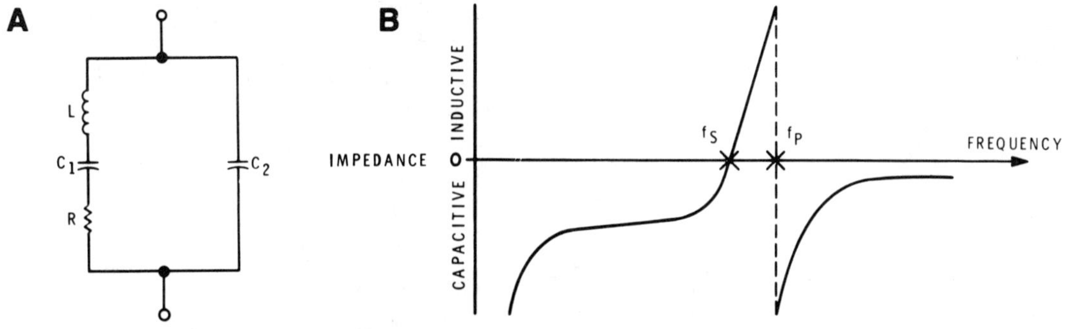

Figure 2-18
(A) Equivalent circuit of a quartz crystal,
(B) impedance of the crystal versus frequency.

A **crystal lattice** filter is shown in Figure 2-19. This filter uses two matched pairs of crystals in a special bridge network. In this circuit, X_{1A} and X_{1B} are identical. X_{2A} and X_{2B} are also identical to each other but slightly different from X_{1A}–X_{1B}. The impedance diagram for the crystals is shown in Figure 2-20. Note that there are two parallel resonant frequencies for each crystal. This is due to the inductance present in the circuit. They will form resonant circuits with the crystals at specific frequencies. Let's analyze this circuit in detail.

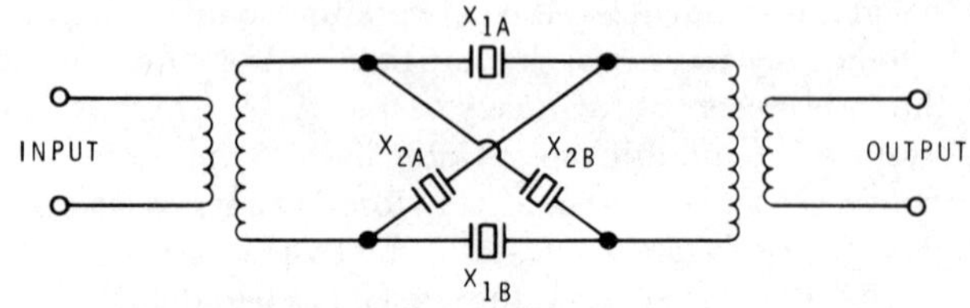

Figure 2-19
A crystal lattice filter.

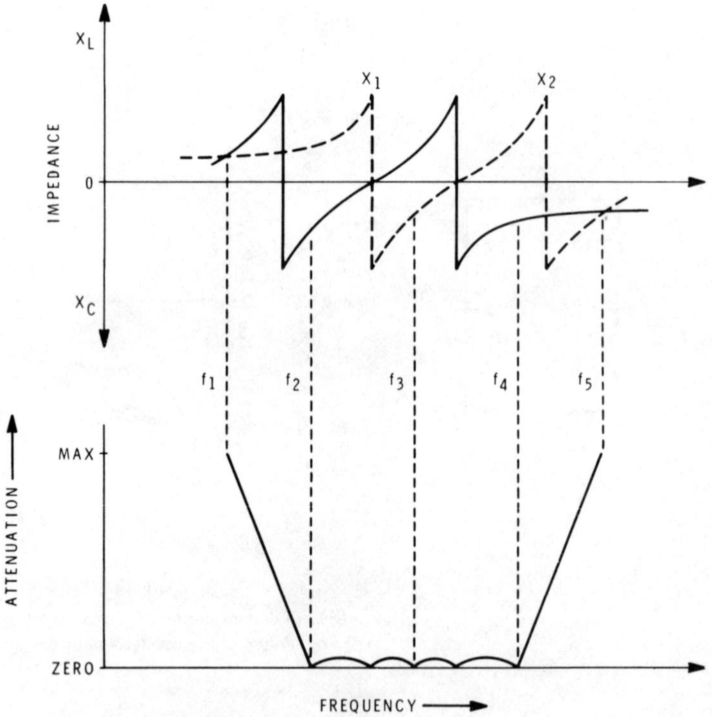

Figure 2-20
Impedance diagram and selectivity curve of a crystal lattice filter.

At f_1 in Figure 2-20, both X_1 and X_2 present equal impedances to the input signal. Therefore, in the circuit of Figure 2-19, equal amplitude signals are passed through X_{1A} and X_{2A}. However, since they are from opposite sides of the transformer, they are 180° out of phase, and cancel. The same holds true for X_{1B} and X_{2B}. This is a point of maximum attenuation for the filter as shown on the response curve of Figure 2-20.

At f_2, both X_1 and X_2 have equal impedances again. However, X_1 acts inductively while X_2 acts capacitively. The result is that the signals arrive at the output in phase. For example, if the signal at X_{1A} is at 0°, since X_{1A} acts inductively, it will shift the signal +90°. The signal at X_{2A} is at +180° due to the transformer. X_{2A} acts capacitively and, therefore, shifts the signal −90°. The result is that both signals arrive at the output with a phase angle of +90°. They are in phase and there will be zero attenuation at this frequency. This is shown in Figure 2-20.

At a frequency midway between f_2 and f_3, X_2 attains series resonance while X_1 is in parallel resonance. The signal, therefore, passes unattenuated to the output through X_{2A} and X_{2B}. At f_3, the crystal's impedances are again equal but opposite and zero attenuation occurs. Midway between f_3 and f_4 crystal X_1 reaches series resonance and yet another zero attenuation point occurs. At f_4 again, the impedances are equal but opposite and zero attenuation occurs. However, at f_5, the impedances are equal and in phase. The output signals, therefore, cancel and maximum attenuation occurs.

As you can see, the crystal lattice filter makes an excellent sideband filter with relatively steep attenuation curves or "skirts" on both sides of the passband. However, for most applications, even steeper skirts and more attenuation of the unwanted sideband are required. For this reason, several lattice sections must be cascaded to achieve the required response.

Another type of filter used extensively in single sideband transmitters is the **mechanical filter**. A typical mechanical filter is shown in Figure 2-21. The coupling coil along with the biasing magnet form an electrical-to-mechanical or magneto-striction transducer. Any electrical signal across the coupling coil will set up a magnetic field which will either oppose or attract the biasing magnet. This magnetic action will cause the coupling coil and transducer rod to move. This mechanical motion will then be coupled to the disc resonators. These discs are cut to precise dimensions and, therefore, will resonate at only a certain band of frequencies. When the discs resonate, they will pass the signal on to the output transducer. If they do not resonate, they will attenuate the input signal substantially.

Mechanical filters offer excellent characteristics and are often chosen for the most exacting applications. However, they are limited to operation in the 60 – 600 kHz range, whereas crystal filters can be constructed up to 10 MHz and higher.

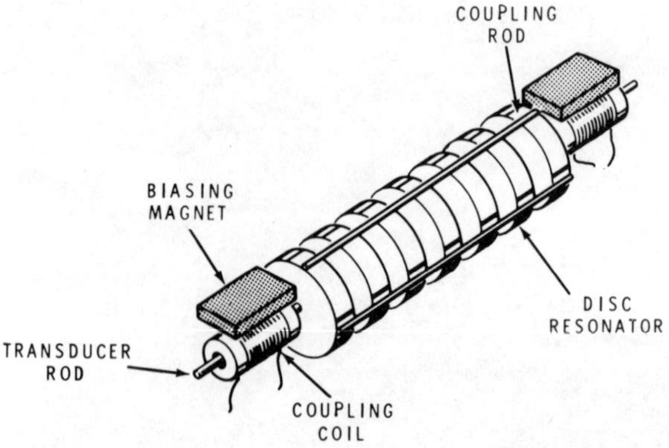

Figure 2-21
A sideband filter using mechanical disc resonators.

FREQUENCY CONVERSION

Thus far in the SSB transmitter we have an upper or lower sideband signal present at the filter's output. However, this is a fixed frequency signal. That is, the sideband filter is fixed-tuned and, therefore, the SSB signal is also restricted to a single frequency. We must now convert this SSB signal to the desired operating frequency.

Figure 2-22 shows a typical frequency conversion technique. It uses a mixer and a heterodyne oscillator. The mixer is any nonlinear circuit and could even be a balanced modulator. The heterodyne oscillator is a crystal oscillator, a variable frequency oscillator, or a combination of both. Let's examine Figure 2-22 in detail.

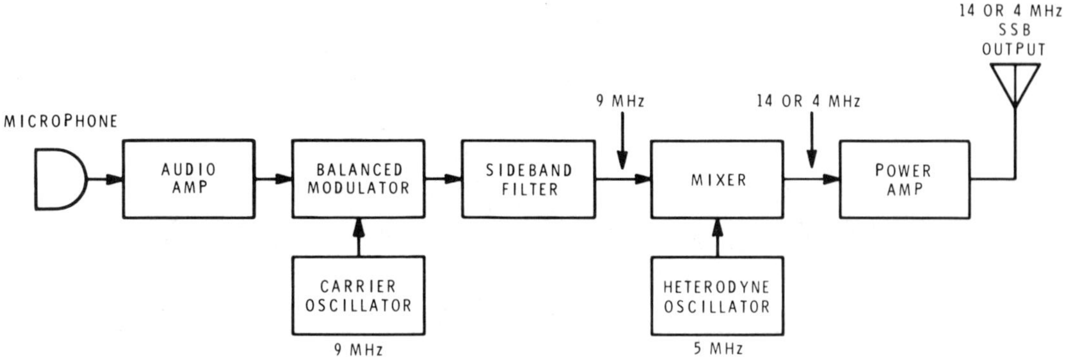

Figure 2-22
Complete SSB filter transmitter.

The carrier oscillator is crystal controlled at 9 MHz which is the passband of the sideband filter. Therefore, the SSB signal, either upper or lower sideband, is applied to the mixer at an approximate frequency of 9 MHz. The heterodyne oscillator frequency for this example is 5 MHz, although it may be switched or varied to obtain any desired output frequency. In this case, the 9 MHz and 5 MHz signals are applied to the mixer and the result is sum and difference mixing products. Therefore, either 4 MHz or 14 MHz will appear at the output. The desired mixing product is selected by a resonant tank circuit at the output of the mixer. The power amplifier is also tuned to the desired frequency. It amplifies the low-level SSB signal to the needed level for transmission. Thus, once the SSB signal is generated at a low-level fixed frequency, it is converted to the desired operating frequency using a heterodyne oscillator and a mixer.

SIDEBAND SELECTION

There are three methods of selecting either the upper or the lower sideband. The obvious method, using two different filters, is shown in Figure 2-23A. The only drawback to this method is the additional cost of the second filter. The most commonly used technique is shown in Figure 2-23B. This method shifts the frequency of the carrier oscillator. For upper sideband, the carrier oscillator frequency is set to the lower edge of the filter's passband. For lower sideband, it is set to the upper edge of the passband. The carrier frequency is usually changed by switching between two crystals.

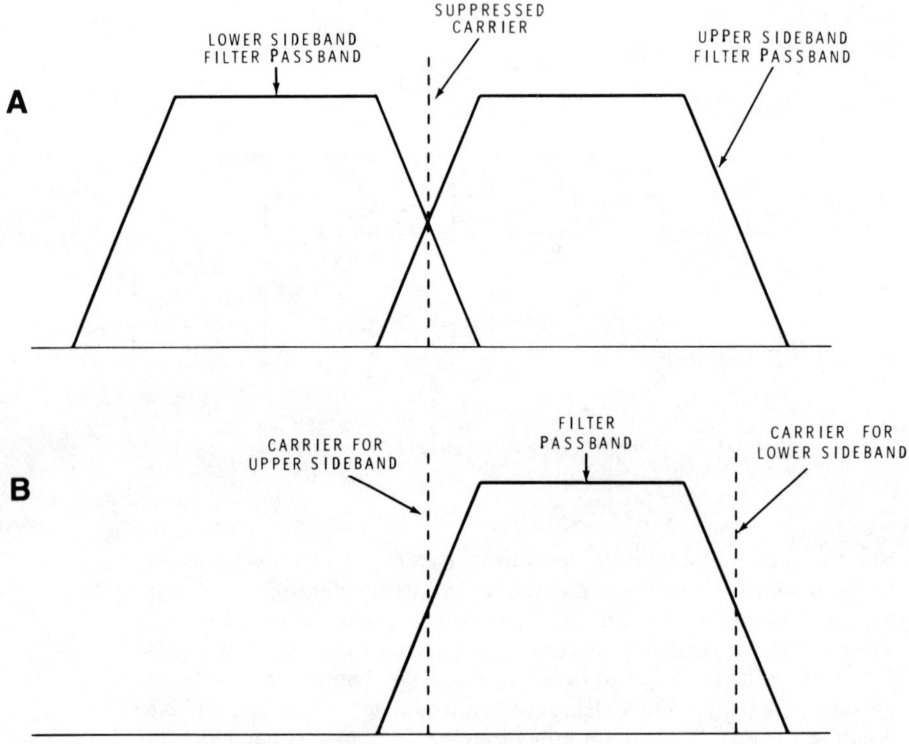

Figure 2-23
Two methods of sideband selection.

The third method of sideband selection involves frequency inversion. This can occur during the heterodyne or frequency conversion process. As an example, suppose an upper sideband signal is generated with a suppressed carrier of 9 MHz and sideband frequencies extending to 9.003 MHz. When this signal is mixed with a 15 MHz heterodyne oscillator signal, the sum products will be a 24 MHz suppressed carrier with sideband components out to 24.003 MHz. This is still an upper sideband signal. However, the difference products will be a 6 MHz suppressed carrier and sideband components down to 5.997 MHz. This is now a lower sideband signal.

Frequency inversion can be used to select the desired sideband or one of the other two methods. However, whichever method is chosen, the effects of frequency inversion must be taken into account when designing the transmitter.

Phasing Method of SSB

The only disadvantage of the filter method is the expensive sideband filter required. To get around this, the phasing method of generating a single sideband signal was developed. The block diagram for this system is shown in Figure 2-24.

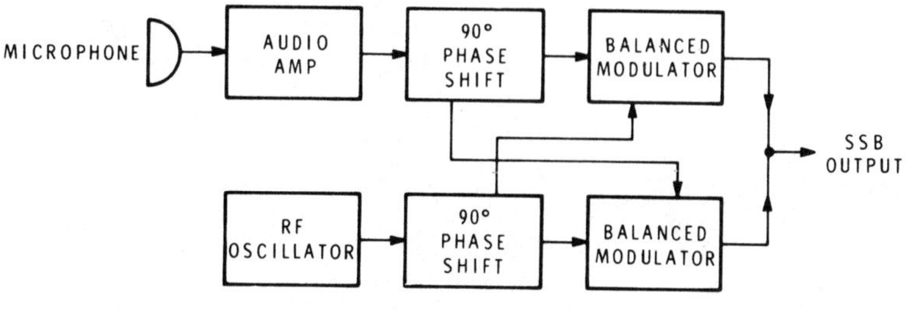

Figure 2-24
Phasing method SSB transmitter.

With this system, a 90° phase shift network is used to split the audio signal into two components that have a 90° phase difference. The RF oscillator signal is also split into two components that have a 90° phase difference. One RF and one audio signal are then applied to each of the two balanced modulators.

Both balanced modulators suppress the carrier and have a double sideband output. These output signals are then added together and, due to the phasing of the applied signals, one sideband is cancelled and the other sideband is reinforced.

The disadvantage of the phasing method is that it requires considerable and precise adjustment to operate correctly. For this reason, the filter method is used exclusively in modern single sideband transmitters.

Linear Amplifiers

A single sideband signal is usually generated at a very low power level, often less than one watt. To increase its power level, one or more power amplifiers must be used. However, these power amplifiers must be linear; they must amplify the SSB signal without introducing any distortion. Otherwise harmonics and other forms of "spurious radiation" will appear at the output and cause interference.

For linear amplification of an SSB signal, the amplifier must be biased for class A, AB, or B operation. Class AB and B can be used because the flywheel effect of the output tank will complete the RF waveform. However, class C amplifiers will cause serious distortion and cannot be used. Normally class AB or B amplifiers are used because they offer the best efficiency.

SSB Power Ratings

The power rating for an AM transmitter is its DC input power under "carrier only" conditions, or in other words, with no modulating signal. However, a single sideband transmitter does not have a carrier; in fact, with no modulating signal, there is no output. This makes power measurements slightly more difficult with an SSB transmitter.

Figure 2-25 shows the output waveform of an SSB transmitter with a normal speech modulating signal. The power contained in the maximum peak of the signal is the basic transmitter rating and is called the **peak envelope power** (PEP). However, envelope peaks occur only sporadically during voice transmission. Therefore, an average power reading is much easier to obtain. In fact, the final amplifier current and voltage meters respond only to the average value. This is because the needles of the meters are unable to respond to the rapid peaks and thus give an average indication.

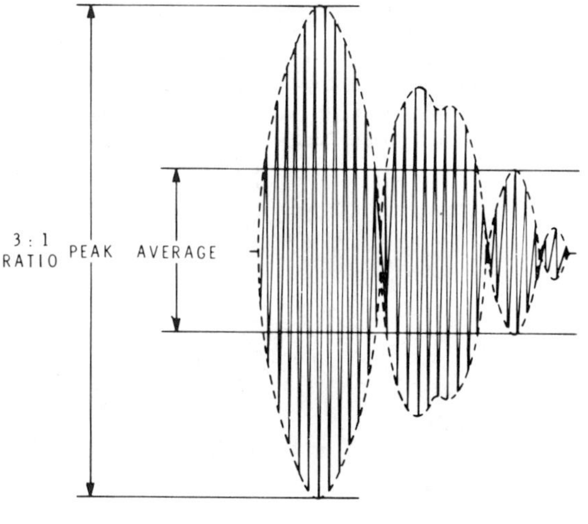

Figure 2-25
SSB output waveform.

To determine the PEP input from the average meter readings, you must determine the ratio between peak and average. This can be very difficult to do in the case of voice modulation because each person's voice characteristics are different. However, a two-tone test signal may be used that consists of two audio sine waves with a frequency separation of approximately 1,000 Hz. This test signal is then applied to the input of the SSB transmitter and the resulting output waveform will appear as shown in Figure 2-26.

With a two-tone test signal, the ratio between PEP input and average power input is 2:1. Therefore, the PEP rating is found by multiplying the input power reading (I × E) by 2. The ratio for a voice-modulating signal may be anywhere in the range of 1.5:1 to 6 or 7:1. Therefore, the two-tone test signal gives an accurate indication of the transmitter's PEP rating.

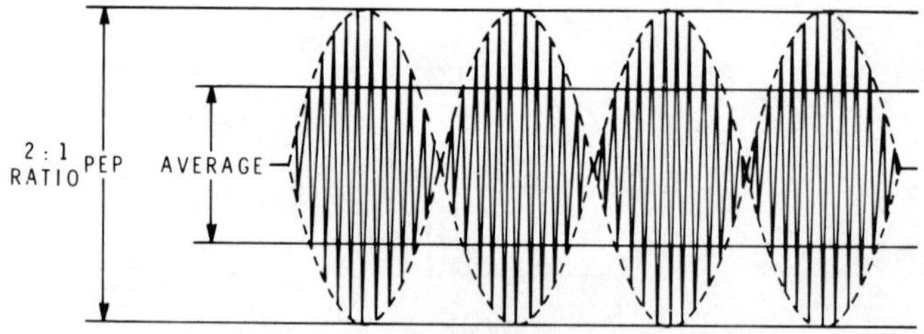

Figure 2-26
Two-tone SSB output.

Unit 3

AM RECEIVERS

INTRODUCTION

The original "wireless" receivers consisted solely of a simple detector. However, with more and more transmitters operating, the receiver had to select one signal from the many that were on the air. This required one or more tuned circuits in conjunction with the detector. Eventually, receivers were required to pick up signals from greater distances, so one or more tuned RF amplifiers were added to the detector to improve sensitivity and selectivity.

However, the tuned RF receiver had several problems. So in 1918, the **superheterodyne** receiver was invented by Edwin Armstrong, who also developed FM broadcasting. The superheterodyne receiver revolutionized the field of radio communications, allowing truly long distance reception for the first time. Even today, virtually all receivers, including TV receivers, use the superheterodyne technique.

In this unit, you will study the basic types of AM detectors, the tuned RF receiver, and the superheterodyne receiver.

AM DETECTORS

As shown earlier, an audio signal is impressed onto a carrier wave in the form of amplitude variations. It is then amplified and applied to a transmitting antenna. As will be shown in a later unit, this modulated signal is then radiated and propagated, and a small fraction of it is collected by the receiving antenna. The receiver must amplify this extremely weak signal and, since the signal is one of many collected by the antenna, the receiver must select the desired signal while rejecting all others. Finally, since modulation took place in the transmitter, **demodulation** must be performed in the receiver to recover the original modulating signal. The circuit that performs this function is called a demodulator or a detector.

The Diode Detector

The most popular AM demodulator is the diode detector. This circuit is very simple and is used in virtually all AM receivers. Its purpose is to recover the envelope from the AM waveform.

The diode detector is shown in Figure 3-1. The switch, S_1, is included merely for explanation. The input to the circuit is the AM waveform that has been selected and amplified by previous stages in the receiver. It is applied to diode D_1, which acts as a half-wave rectifier. The positive half cycles cause D_1 to conduct, developing positive pulses across R_1. D_1 cuts off the negative half cycles of the RF input. The center waveform shows the voltage developed across R_1 if S_1 is open.

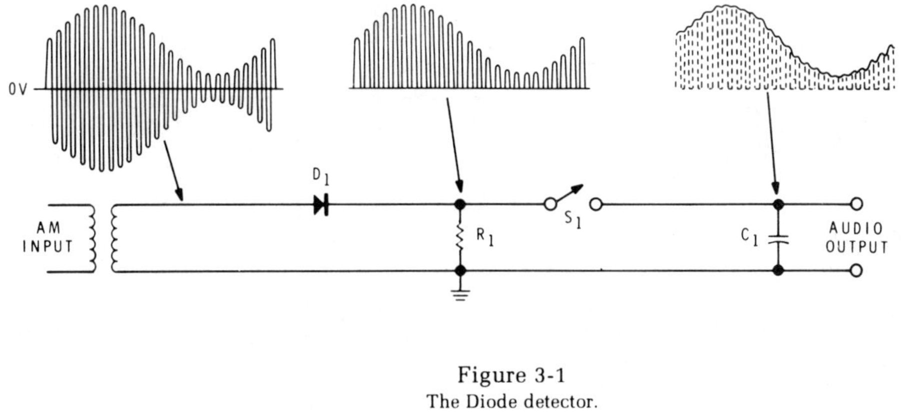

Figure 3-1
The Diode detector.

When S_1 is closed, C_1 is placed in parallel with R_1. C_1 quickly charges through D_1 to the peak of each positive pulse. Between pulses, C_1 attempts to discharge through R_1. However, the RC time constant is chosen so that C_1 discharges only slightly. The result is that the voltage across C_1 follows the envelope of the AM waveform. Thus, the output looks like the upper envelope with a small amount of ripple. Normally, the carrier frequency is many times higher than the envelope frequency, and therefore, the ripple is not noticeable.

Figure 3-1 illustrates the diode detector's operation in the time domain. Let's analyze its operation in the frequency domain.

The AM input consists of three frequency components: the carrier, the upper sideband, and the lower sideband. These signals are applied to D_1 and are mixed across its nonlinear resistance. The difference signal is the modulating information. The next step is to separate this low frequency signal from the high frequency RF. This is accomplished by C_1 which acts as a short circuit to ground for the RF signals and a high reactance for the audio signals.

The final step in AM detection is to separate the audio from the DC component. This is done quite simply using a coupling capacitor such as C_2 in Figure 3-2.

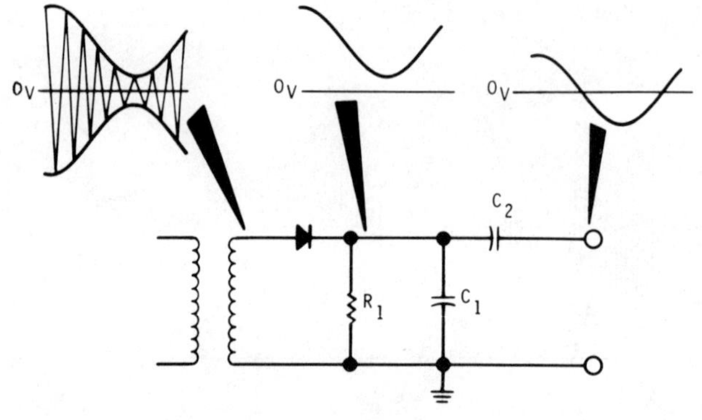

Figure 3-2
Complete diode detector circuit.

The Transistor Detector

Many low cost transistor receivers use a germanium diode as the detector. However, due to the limited RF gain usually preceding the detector, a transistor is sometimes used as the demodulator to provide additional gain. A transistor can perform as a detector, if it is biased for class B operation. In this way, the AM signal is both rectified and amplified at the same time.

Figure 3-3 shows a typical transistor detector. Resistors R_1 and R_2 form a biasing network which sets the circuit for exactly class B operation. R_3 is the collector load resistor, while C_1 filters out the RF components. This leaves the audio signal to be coupled to the output by C_2.

The transistor detector offers a method of achieving additional receiver gain. However, in anything but very low cost receivers, the diode detector is chosen for its simplicity and excellent performance.

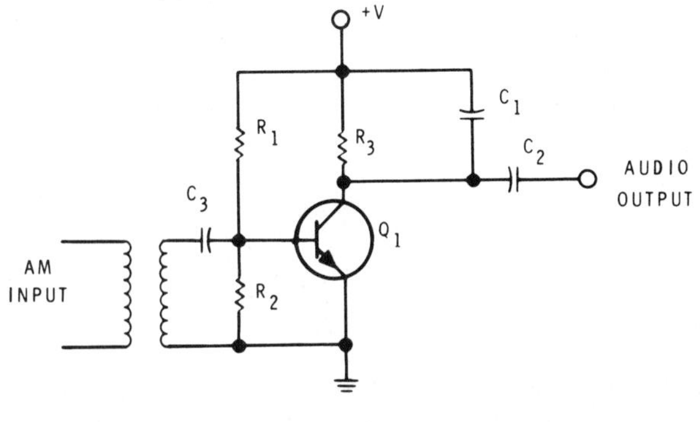

Figure 3-3
The transistor detector.

The Synchronous Detector

The diode and transistor detectors recovered the modulation envelope from the RF signal by rectification. However, with double sideband, suppressed-carrier AM, the envelope does not duplicate the modulating signal. This is shown in Figure 3-4 for a 10 kHz modulating tone and a 100 kHz carrier. Due to the absence of the carrier, the two sidebands combine to produce the envelope shown in Figure 3-4D. This envelope actually has a frequency of twice the modulating frequency or 20 kHz. You can see that this signal could not possibly be demodulated using a diode detector. The recovered signal would have severe distortion and be at twice the frequency of the original modulating information. Therefore, a special circuit called a **synchronous detector** must be used to demodulate a suppressed carrier AM signal.

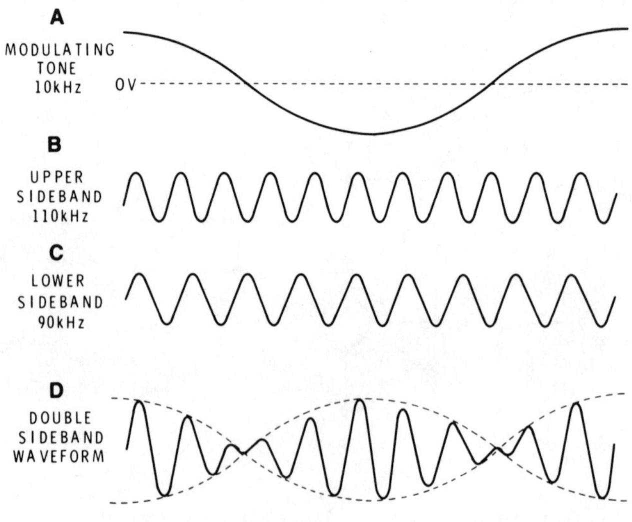

Figure 3-4
When the carrier is removed, the sidebands combine as shown above.

A typical synchronous detector is shown in Figure 3-5. It consists of two diodes rather than the single diode of the simple detector. It also has a separate "carrier" input. This is a signal that is generated by the receiver. As will be seen later, this "carrier" signal must not only be the same frequency as the original suppressed carrier but must also be exactly in phase with the original.

In Figure 3-5, the carrier input causes diodes D_1 and D_2 to switch. When the carrier input results in a negative potential at the top of T_1 and a positive potential at the bottom, both D_1 and D_2 conduct. However, no current flows to the output or C_4 because all current remains in the D_1, D_2, and T_1 loop. In effect, the carrier input is balanced and, therefore does not reach the output. On the other half cycle of the carrier, D_1 and D_2 are reverse biased and no current flows. The combinations of R_1C_1 and R_2C_2 form diode loads and ensure that the diodes are switched on for only a small fraction of each cycle. C_4 filters out any RF components that reach the output.

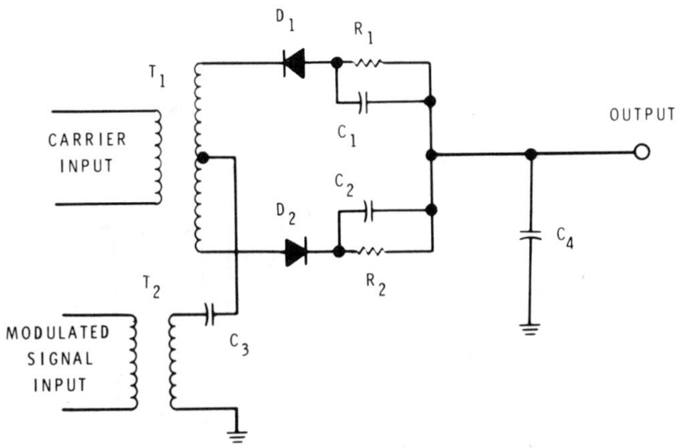

Figure 3-5
A synchronous detector.

The modulated signal input is of a sufficiently low level that it cannot forward bias D_1 and D_2. This means that the only time the modulated signal can reach C_4 and the output is when the carrier input forward biases the diodes. What actually happens is that the carrier input samples small portions of the modulated signal as shown in Figure 3-6. Here, everytime the carrier input goes positive, the diodes are turned on and the suppressed carrier signal passes to the output. Note that the carrier samples the positive peaks during one half cycle of the input and the negative peaks during the other half cycle. The result is an accurate duplication of the modulating signal at the output.

You can see from the waveforms of Figure 3-6 that any slight carrier phase or frequency variation will shift the sampling point and cause severe distortion. For this reason, the carrier signal must be held exactly in phase with the original carrier. This presents the most serious drawback to suppressed-carrier AM. Elaborate circuits are required at the receiver to generate the carrier signal in the correct phase and frequency. However, suppressed-carrier AM is used in the stereo FM and color TV broadcasting systems. In these systems a special synchronizing signal is also transmitted. The receiver recovers this synchronizing signal and uses it to generate its own carrier signal in the correct phase and frequency.

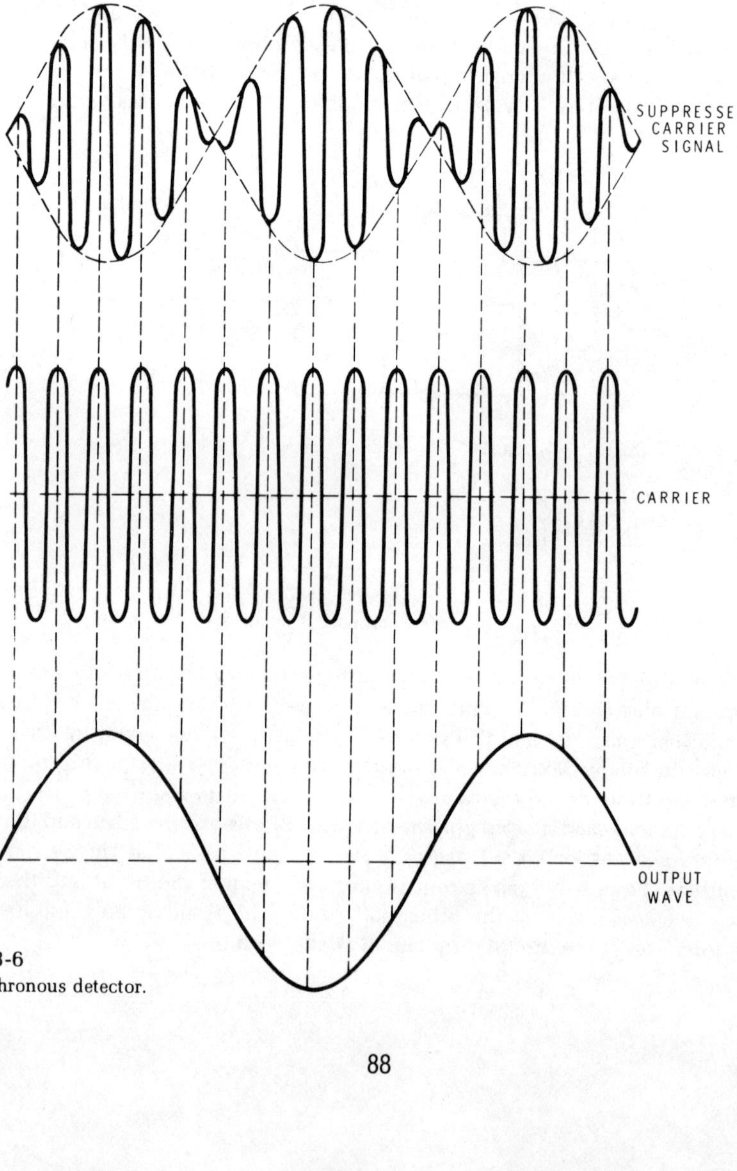

Figure 3-6
Waveforms for the synchronous detector.

The Product Detector

The simple diode detector is also unsuitable for demodulating a single-sideband signal. This is because, with SSB, the waveform's envelope corresponds only to the amplitude of the modulating signal, not the frequency. In fact, if an SSB transmitter is modulated by a constant amplitude audio tone whose frequency is varied, the SSB envelope will be constant but its frequency will vary. Therefore, an SSB demodulator must be sensitive to both frequency and amplitude variations.

In product detection, an internal reference signal is generated and the SSB signal is compared with it. This reference signal is essentially a "substitute" carrier for the single sideband. The product detector itself is nothing more than a nonlinear mixer. It can take on many forms but is usually a balanced modulator circuit or a dual-gate MOSFET.

A typical product detector is shown in Figure 3-7. It uses a diode-ring balanced modulator as the nonlinear mixer. The internally generated carrier, which is also called the BFO for Beat Frequency Oscillator, is applied to the center tap of T_1. The incoming SSB signal is applied to the primary of T_1. The two signals are then mixed across the diode ring. Now, if the carrier signal is within approximately 50 Hz of the original suppressed carrier, the difference signal will be the original modulating audio. $R_1R_2C_1C_2$ form an RC low pass network which removes any RF component and allows only the audio signal to pass to the output. Thus, the original information has been recovered.

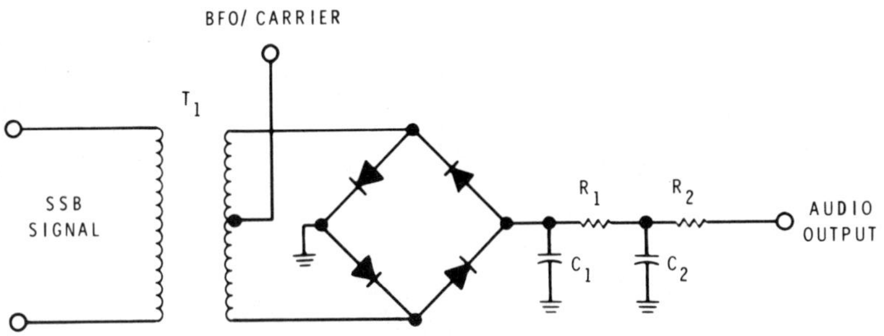

Figure 3-7
Balanced modulator product detector.

As you can see, the requirements for the internally generated carrier for SSB are not nearly as stringent as they are for double-sideband suppressed-carrier AM. In suppressed-carrier AM, the local carrier has to be exactly the same frequency and phase as the original. In SSB, it has to be only within 50 Hz of the original. This greatly simplifies the SSB receiver and, along with bandwidth considerations, is the primary reason for the superiority and widespread use of SSB instead of double-sideband suppressed-carrier AM.

Since the local carrier is not the same frequency and phase as the original, the audio quality of SSB does suffer somewhat. However, the demodulated signal is still intelligible and, therefore, quite effective for voice communications. Lower audio quality is a small price to pay for the many advantages that single-sideband offers.

TUNED RF RECEIVER

If a simple diode detector was connected directly to an antenna, there would be no means of selecting a specific signal. In fact, all signals present at the antenna would be demodulated and the output would be a jumbled mess. One way to select the desired signal is to connect a tuned RF amplifier between the antenna and detector. However, the bandwidth of a single RF amplifier is not narrow enough to reject all the unwanted signals. Therefore, several RF amplifiers must be used, as shown in Figure 3-8. This decreases the bandwidth to a point where only one signal will pass to the detector. The RF amplifiers will also increase the amplitude of the desired signal before it is applied to the detector. Thus, the additional RF amplifiers increase both the sensitivity and selectivity of the receiver.

Sensitivity is a measurement of a receiver's ability to receive weak signals. For example, if a receiver can "pick up" an extremely weak signal, it is said to have good sensitivity. The amount of gain a receiver has determines its sensitivity. Therefore, the more gain obtained in the RF section, the greater the receiver's sensitivity is.

Selectivity is a measure of a receiver's ability to select one signal while rejecting all others. In a tuned RF receiver, this is determined by the bandwidth of the RF amplifiers. Therefore, as the bandwidth is reduced, selectivity is increased.

The receiver shown in Figure 3-8 has four RF amplifier stages. Its sensitivity and selectivity are adequate for the AM broadcast band of 535 to 1605 kHz. However, it is extremely difficult to get all the RF amplifiers to tune to exactly the same frequency as the dial is tuned across the band. Another problem is that the bandwidth of the RF amplifiers increases as the operating frequency increases. As a result, the selectivity at 1600 kHz is not as good as it is at 540 kHz. One other problem is the extreme difficulty of designing RF amplifiers that provide high gain and are tunable over the required range. Because of these and other problems, the tuned RF receiver was abandoned in favor of the superheterodyne receiver. The next section discusses the superheterodyne receiver in detail.

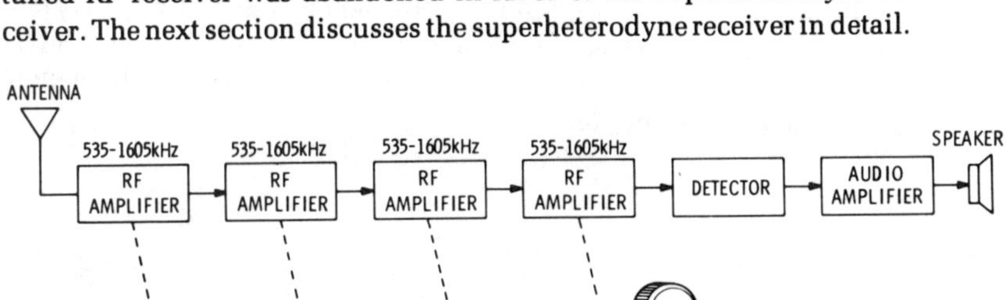

Figure 3-8
The tuned RF receiver.

SUPERHETERODYNE RECEIVERS

Most disadvantages of the tuned RF receiver disappear if the receiver is fixed-tuned to a single frequency. The superheterodyne receiver uses this principle by converting the selected RF signal, regardless of frequency, to a predetermined fixed frequency where amplification and further selection take place.

Block Diagram

The block diagram of a typical, standard AM broadcast superheterodyne receiver is shown in Figure 3-9. The RF amplifier is similar to that used in the tuned RF receiver. It provides signal gain and a limited degree of selection. However, the majority of the receiver's selectivity and gain are obtained from the later stages. In fact, in some receivers, the RF amplifier is eliminated altogether.

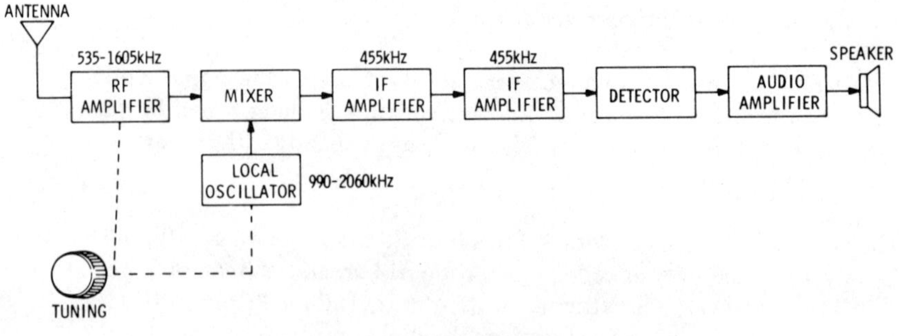

Figure 3-9
Standard AM broadcast superheterodyne receiver.

The mixer and local oscillator are the key to the superheterodyne action. They convert the selected RF signal to a fixed frequency for further amplification. In this case, the fixed frequency is 455 kHz which is a commonly used **intermediate frequency** or IF. It is called an "intermediate" frequency because it is between the original radio frequency and the detected audio frequency.

To convert the incoming RF signal to 455 kHz it must be mixed or heterodyned with a signal that is exactly 455 kHz above or below it. That is, the difference between the selected RF signal and the local oscillator must be 455 kHz. In the receiver of Figure 3-9 the local oscillator is tuned above the incoming signal. Therefore, if the desired signal is at 1000 kHz, the tuning knob would be adjusted to 1000 kHz. This would adjust the RF amplifier to 1000 kHz and allow a relatively small band of frequencies to be amplified and applied to the mixer. The tuning knob would also adjust the local oscillator to exactly 455 kHz above the incoming RF signal or 1000 kHz + 455 kHz = 1455 kHz.

Both the RF and local oscillator signals are then applied to the mixer. The mixer is any nonlinear circuit and could even be a balanced modulator. The outputs of the mixer are the original signals and the sum and difference mixing products. Since the mixer's output and several following amplifiers are fixed-tuned to 455 kHz, only the difference signal, in this example 1455 kHz − 1000 kHz or 455 kHz, leaves the mixer and is amplified.

The IF amplifiers are actually fixed-tuned RF amplifiers. There are usually tuned resonant circuits on both the input and output of each stage. This assures both high gain and narrow bandwidth which means better receiver sensitivity and selectivity. Since they are fixed-tuned, their gain and bandwidth are constant. There are also no tuning problems since each stage is individually "peaked" only once, during its initial alignment at the factory. The selection of the intermediate frequency itself involves several factors which are discussed later in this unit.

The amplified IF signal is then coupled to the detector where the original modulating information is recovered. In the standard AM receiver of Figure 3-9, the detector is a simple diode demodulator. A single-sideband receiver would require a product detector and a BFO. The detected audio is then amplified and applied to the speaker.

Frequency Conversion Circuits

A typical receiver front-end circuit is shown in Figure 3-10. Q_1 is the RF amplifier, Q_2 is the mixer, and Q_3 is the local oscillator. The RF amplifier is tuned to the desired input frequency by C_1. As shown in the schematic, C_1 is also mechanically coupled to C_2. Therefore, C_2 simultaneously tunes the local oscillator to its correct frequency. C_2 resonates with L_1 to form the tank for the oscillator while L_2 provides the regenerative feedback.

L_3 couples the local oscillator signal to the emitter of Q_2, while the input RF signal is coupled to the base. These two signals are then mixed across the nonlinear resistance of Q_2. The output of the mixer is applied to the primary of T_3 which is tuned to the intermediate frequency. Thus, the difference signal, which is the intermediate frequency, is separated and applied to the IF stages where it is further amplified.

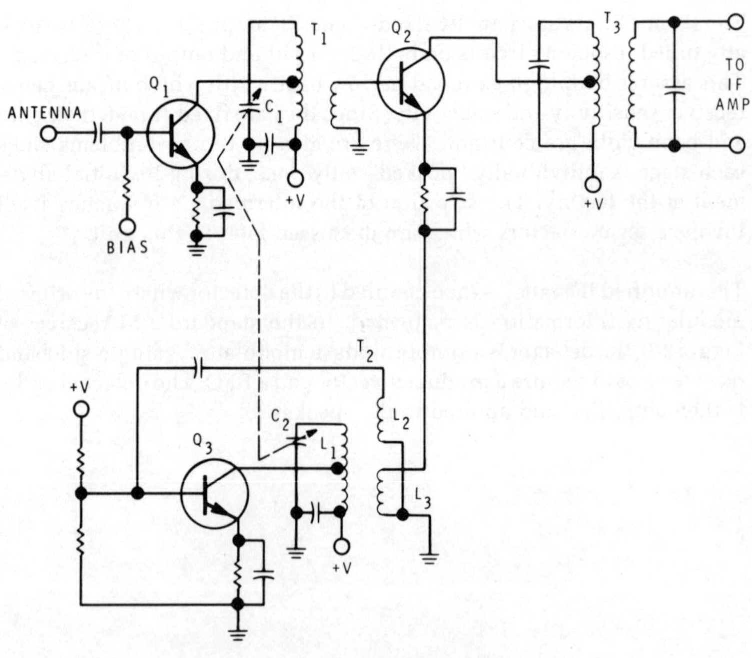

Figure 3-10
Frequency conversion circuit.

A variation of the previous circuit is shown in Figure 3-11. However, in this circuit, transistor Q_2 performs the dual function of local oscillator and mixer. The key to this circuit is transformer T_2. Here, L_1 couples the collector signal of Q_2 to the resonant tank of L_2 and C_2. C_3 then couples the oscillations of L_2C_2 to the emitter of Q_2. The transistor amplifies this signal and the continued feedback of L_1 sustains the oscillations. Thus, Q_2 functions as an Armstrong oscillator.

When a signal is present at the antenna, Q_1 and the tuned circuit of C_1T_1 select and amplify the desired signal. It is then coupled to the base of Q_2 and mixed with the local oscillator signal. The output of Q_2 is coupled through L_1 to the primary of T_3 which is tuned to the intermediate frequency. Thus, the difference signal is separated and applied to the IF stages. This type of circuit is called an **autodyne**. It is usually used in low cost portable receivers, and is often used in conjuction with the transistor detector discussed earlier. Both circuits perform dual functions and, thus, allow simplicity and lower cost.

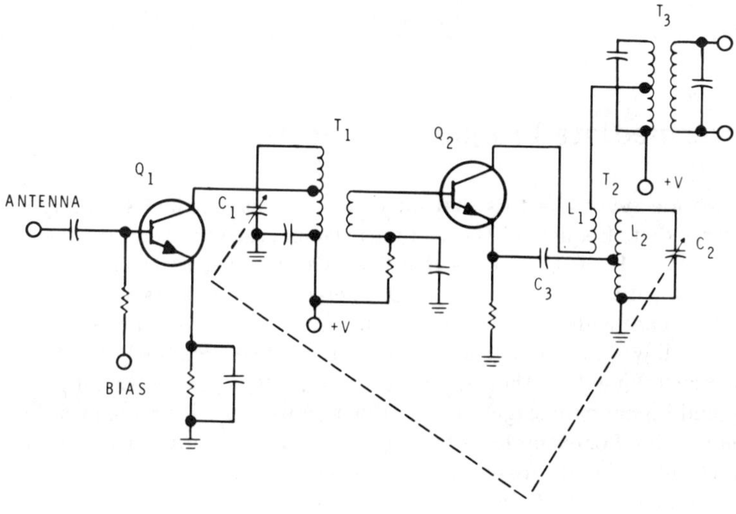

Figure 3-11
Autodyne conversion circuit.

A high performance mixer circuit often found on HF, VHF, and UHF receivers is shown in Figure 3-12. It uses a dual-gate MOSFET as the active element. The MOSFET has a high input impedance and, therefore, causes less loading on the previous circuit. In this mixer the RF signal is applied to gate 2 while the local oscillator signal is applied to gate 1. This minimizes any interaction between the two signals. The mixing action is produced by biasing the MOSFET into its nonlinear region. The difference output signal is then separated by the tuned primary of T_2.

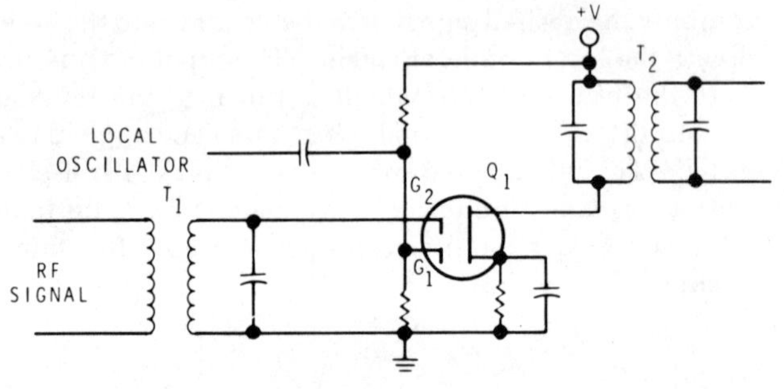

Figure 3-12
A MOSFET mixer.

Intermediate Frequency Selection

There are several factors that must be considered when selecting an intermediate frequency. Many of these factors revolve around the equation $BW = F_0/Q$. This states that bandwidth is equal to the operating frequency divided by the circuit Q. Therefore, for good selectivity, or a narrow bandwidth, the circuit Q must be high and the operating frequency low. Thus, it is generally desirable to use a relatively low intermediate frequency. Other advantages of a low IF are better circuit stability and higher gain stages. This is due to fewer circuit problems at low frequencies. For example, as frequency increases, circuit resistive, radiation, and dielectric losses increase; stray capacitance and inductance effects increase; and stray feedback causes instability. Therefore, low intermediate frequencies are desirable.

Another factor to be considered when selecting an intermediate frequency is the required bandwidth. For example, if an AM broadcast receiver uses a 60 kHz IF with a Q of 60, the bandwidth will be 1 kHz. This is hardly practical for receiving an AM signal whose bandwidth is at least 10 kHz! Therefore, the system bandwidth must be taken into account.

The final factor to be considered is **image frequency rejection**. The image frequency is an undesired RF signal which, when mixed with the local oscillator, will also produce the correct IF value. As an example, suppose the IF is 200 kHz and the desired signal is 4.2 MHz. If the local oscillator is tuned above the incoming signal, it must be at 4.2 MHz + 0.2 MHz = 4.4 MHz. Therefore ,when the local oscillator and RF signal are mixed, the difference signal will be 200 kHz which is the IF. However, there is another signal that will produce a 200 kHz IF signal. It is located 200 kHz above the local oscillator at 4.6 MHz. Therefore, when this 4.6 MHz signal is mixed with the 4.4 MHz local oscillator the difference is 200 kHz. Thus, this image frequency will be amplified just as if it were the desired signal.

Let's try another example of finding the image frequency. If the IF is 455 kHz and the received signal is 1110 kHz with the local oscillator tuned above the RF signal, what is the image frequency? The local oscillator frequency is 1110 kHz + 455 kHz = 1565 kHz. Therefore, the image frequency is 1565 kHz + 455 kHz = 2020 kHz.

From the previous examples, you can see that the image frequency is separated from the desired signal by twice the intermediate frequency. Rejection of the image signal depends solely on the selectivity of the RF amplifier and mixer stages. Also, the higher the IF is, the further the image signal will be from the desired signal. Therefore, to improve image rejection, a higher intermediate frequency should be used. This is in conflict with the previous factors considered. Therefore, the selection of an intermediate frequency must be a compromise.

Once all of these factors were taken into account, several "standard" intermediate frequencies were chosen. As an example of this, AM broadcast receivers (535 – 1605 kHz) generally use an IF of 455 kHz. This frequency provides excellent Q for high gain and good selectivity. It is also high enough to place the image frequency well outside the RF amplifier's passband. FM broadcast receivers use an IF of 10.7 MHz. This was chosen to provide sufficient bandwidth, up to 200 kHz, for the FM signal. It also places the image frequency 21.4 MHz away from the desired RF signal. Television receivers use an IF of approximately 40 MHz. This allows the extremely wide TV bandwidth of 4 MHz to be accomodated.

IF Amplifiers

An IF amplifer is actually nothing more than a fixed tuned RF amplifier. Figure 3-13 shows a typical amplifier tuned to 455 kHz. The only difference between this circuit and a typical RF amplifier is the double-tuned transformers T_1 and T_2. The resonant circuits in both primary and secondary improve the selectivity and circuit gain. Apart from this, the circuit is a straightforward RF amplifier with C_1 providing neutralization for added stability.

The selectivity of a receiver can be tailored to the exact requirements using a bandpass filter as part of the IF section. This technique is used primarily in communications receivers where narrow bandwidths are used, in FM receivers where the high 10.7 MHz IF does not provide adequate selectivity, and it is starting to appear in TV receivers where it eliminates tedious tuning and provides optimum IF response.

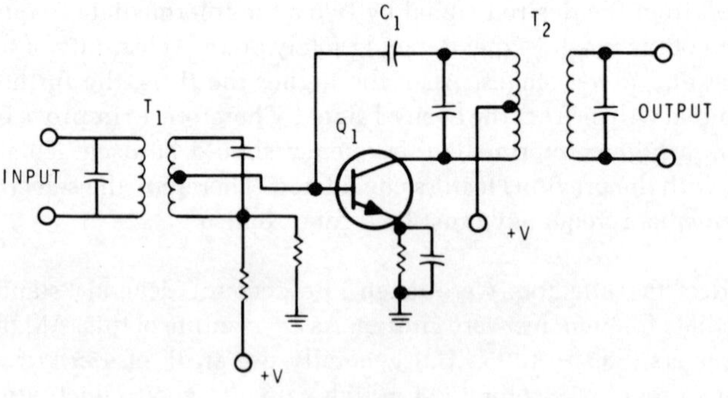

Figure 3-13
An IF amplifier.

Automatic Gain Control

When you tune across the AM Broadcast band, it is not uncommon to hear an extremely weak station at one point and a strong local station at the next frequency. To compensate for these level variations, Edwin Armstrong invented automatic volume or gain control. This circuit automatically varies the gain of the receiver to keep the volume at a relatively constant level regardless of the input signal.

A simple automatic gain control (AGC) circuit is shown in Figure 3-14. Actually it is an AM diode detector with R_2 and C_2 added. These components take the detected and filtered audio signal and filter it still further to remove the audio signal. What remains is a DC voltage that is directly related to the input signal amplitude. If a strong input signal is present, the AGC voltage will be high. Just the opposite is true for a weak signal. This voltage is then used to control the receiver gain. While a positive AGC voltage is shown in Figure 3-14, negative voltage is also used and is obtained by reversing the connection of D_1.

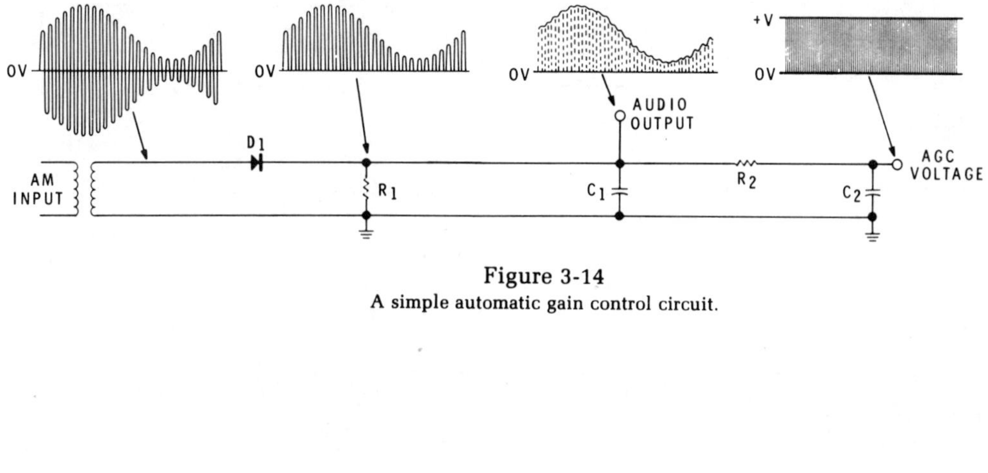

Figure 3-14
A simple automatic gain control circuit.

To see how the AGC voltage controls receiver gain, consider the block diagram shown in Figure 3-15. Here, the DC voltage derived from the detector is applied to both IF amplifiers and the RF amplifier. This voltage is used to set the bias of these stages. Therefore, by varying the stage bias, the gain of each stage can be controlled. For example, when a strong signal is present, the AGC voltage is high and it decreases the gain of the RF and IF amplifiers. If the signal fades or a weak signal is tuned in, the AGC voltage is reduced and the RF and IF amplifier's gain is increased. The result is a relatively constant audio output level.

You can see that the AGC circuit is a valuable addition to a receiver. However, some modes of operation such as SSB do not have constant signals as with AM. This makes it difficult to derive an AGC voltage. In this case, special AGC circuits have been developed. Although, in some cases, the AGC circuit is turned off altogether and a manual "RF gain" control is used.

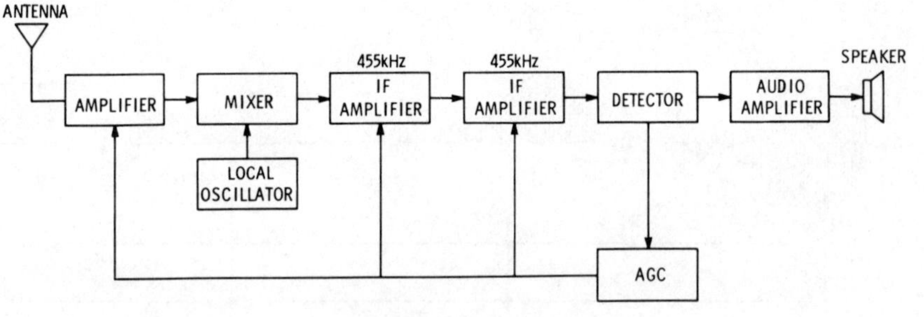

Figure 3-15
A superheterodyne receiver with AGC.

An AGC circuit used with SSB receivers is shown in Figure 3-16. Since there is no carrier present in SSB, the AGC system must derive its voltage from some other part of the signal. The circuit of Figure 3-16 uses the demodulated audio to obtain the AGC voltage. Here, the audio is amplified by the operational amplifier and then rectified by the voltage doubler circuit of D_1D_2 and C_1C_2. Therefore, the DC voltage across C_2 is directly proportional to the strength of the SSB signal. However, if no audio is transmitted this voltage drops to zero. In fact, in between words and sentences, when the speaker pauses, C_2 attempts to discharge through R_1. However, the time constant of C_2R_1 is sufficient to prevent the AGC voltage from dropping during normal pauses in speech.

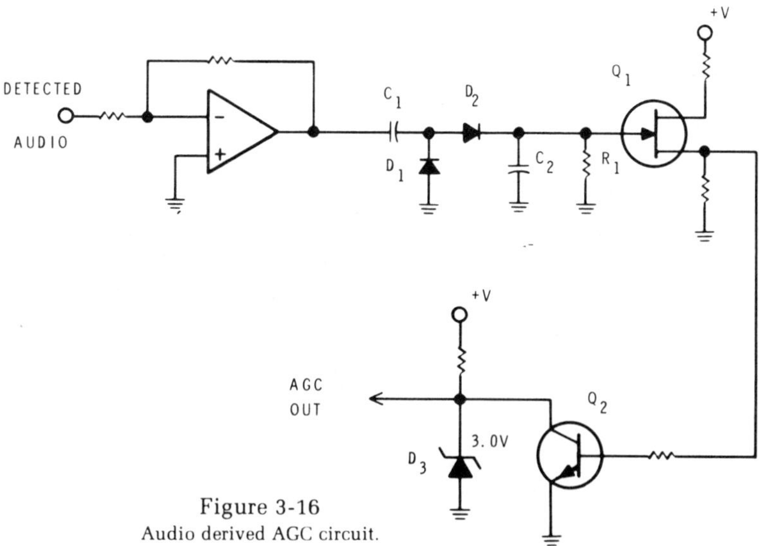

Figure 3-16
Audio derived AGC circuit.

JFET Q_1 and transistor Q_2 form an interface between the AGC detector and the receiver circuits being controlled. Q_1's high input impedance allows it to amplify the DC voltage across C_2R_1 without affecting their time constant. The output of Q_1 is applied to the base of Q_2 and when the voltage is sufficient, it turns on Q_2. Up until this time the AGC voltage is held at a constant 3 V by zener diode D_3. When Q_2 turns on, it effectively turns off the zener by reducing the voltage below 3 V. Q_2 now controls the AGC voltage and any increase in the detected signal at C_2 will cause Q_2 to conduct more, which lowers its collector voltage and, hence, the AGC voltage. Thus, as the received signal increases, the AGC voltage decreases which reduces the receiver gain and holds the audio output relatively constant.

Double Conversion Receivers

As mentioned earlier, choosing an intermediate frequency is usually a compromise. A low IF is desired because of its better gain, stability, and narrow bandwidth. However, a high IF is desirable to improve image rejection. The double-conversion receiver is designed to take advantage of both high and low IFs by using a high "first" IF and then converting to a low "second" IF.

A simple double-conversion receiver is shown in Figure 3-17. This receiver uses two mixers, two local oscillators, and two intermediate frequencies. The first IF is selected to place the image as far away from the desired signal as possible. In this case, the 9 MHz first IF places the image 18 MHz away from the desired signal. Therefore, the image is well outside the bandwidth of the RF amplifier and is greatly attenuated.

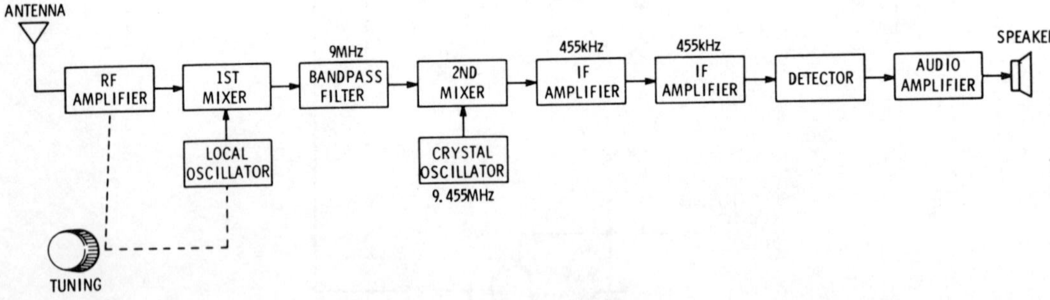

Figure 3-17
Double conversion receiver.

To demonstrate how the receiver operates, we must follow the signal from the antenna to the speaker. As an example, suppose you want to receive the 5 MHz signal of the standard time and frequency station, WWV. The tuning knob is set to 5 MHz which tunes the RF amplifier to 5 MHz and the local oscillator to 14 MHz. This places the image frequency at 23 MHz, well outside the bandpass of the RF amplifier.

The incoming signal is amplified and then applied to the mixer along with the local oscillator signal. At the mixer output, the difference signal of 9 MHz is selected by the bandpass filter and applied to the second mixer. The bandpass filter may be an LC filter or a crystal filter for improved selectivity. In this receiver there is no first IF amplifier since most of the receiver's gain is supplied by the second IF.

Since a 9 MHz signal always appears at the second mixer, a crystal-controlled "second" local oscillator is used. Its frequency of 9.455 MHz is mixed with the 9 MHz input signal and the difference frequency, 455 kHz, is selected by the IF amplifiers.

The IF amplifiers increase the signal amplitude and apply it to the detector. The detector demodulates the signal and recovers the original audio. This audio signal is then amplified and applied to the loudspeaker.

Receiver Specifications

In the 1920's and 30's the most important receiver specification was sensitivity. The receiver had to be sensitive enough to "pick up" the weak signals of the relatively few transmitters that were "on the air." However, in the 1940's and 50's there were many more transmitters, and receiver selectivity became much more important. Today, there are literally millions of transmitters and, while sensitivity and selectivity are still important, many new specifications are recognized as important receiver performance measurements. Let's look at these specifications in detail.

SENSITIVITY

One of the most important receiver specifications is **sensitivity**. It is a measure of the amplitude of the weakest signal that can be received, with a specified signal-to-noise ratio. That is, the signal must be higher than the noise level by a certain ratio.

Receiver sensitivity is actually a measure of how much signal it takes to overcome the internal receiver noise. This is the reason for using a specified signal-to-noise ($\frac{S}{N}$) ratio. In fact, signal-to-noise ratio measurements have been replaced by a more correct measurement of signal-plus-noise-to-noise ($\frac{S+N}{N}$) ratio. This ratio takes into account that the noise is still present, along with the signal. The signal is actually added to the noise and must overcome it.

Receiver sensitivity is determined by first measuring the noise output with no signal present. The output from an RF signal generator is then applied to the receiver input. The generator's output level is now increased until the receiver's output increases by 10 dB. Once this is achieved, the RF signal generator's output voltage is measured. Typical values are 1.0 to 0.5 μV of RF signal for a +10 dB signal-plus-noise-to-noise ($\frac{S+N}{N}$) ratio. Thus, the receiver requires approximately 1.0 μV of signal to overcome the internal noise by 10 dB and to provide adequate reception.

SELECTIVITY

Another important receiver specification is **selectivity**. This is a measure of the receiver's ability to select a desired signal and reject all others. The only way to measure selectivity is to determine the receiver's bandwidth. A typical receiver selectivity curve is shown in Figure 3-18. This shows that the receiver bandwidth is 3 kHz at the −6 dB point. That is, at 1.5 kHz either side of the center of the passband, the received signal's amplitude drops 6 dB. At 3 kHz either side of the center, the signal drops 60 dB. Therefore, the bandwidth at −6 dB is 3 kHz, and at −60 dB, is 6 kHz. This selectivity curve is primarily due to the bandpass filter in the IF section, although the RF and IF amplifiers also play an important part in improving selectivity.

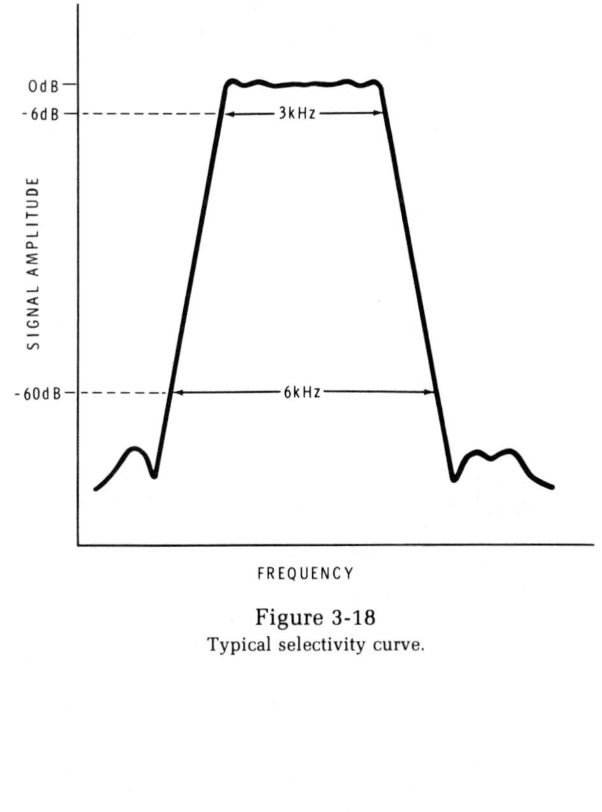

Figure 3-18
Typical selectivity curve.

IMAGE REJECTION

Image rejection is a measure of the receiver's ability to reject a strong signal at the exact image frequency. To measure image rejection, you connect a signal generator to the receiver and set it to the image frequency. For example, if the receiver's first IF is 9 MHz, you would set the receiver to a specific received frequency, let's say 5 MHz. This places the image at 5 MHz + 18 MHz or 23 MHz. So you would set the signal generator to 23 MHz. Then you would continue as though you were making a receiver sensitivity measurement. You would increase the generator's output level until you achieve a 10 dB (S + N)/N ratio. Now, comparing the signal generator's present output level to the normal sensitivity of the receiver will give you the image rejection.

As an example, suppose the receiver's normal sensitivity is 1 μV for a 10 dB (S+N)/N ratio and it takes 100 μV at the image frequency for a 10 dB (S+N)/N ratio. The image rejection is:

$$\text{Image rejection} = 20 \text{ Log } \frac{E_1}{E_2}$$

$$= 20 \text{ Log } \frac{100\ \mu V}{1\ \mu V}$$

$$= 20 \text{ Log } 100$$

$$= 20 \times 2$$

$$= 40 \text{ dB}$$

Thus, in this example, the image rejection is 40 dB. In other words, the image signal will be 40 dB below the desired signal. Image rejection is basically a measure of the selectivity of the RF amplifier.

IF REJECTION

IF rejection is a measure of the receiver's ability to reject a strong input signal at its own intermediate frequency. For example, if the receiver's IF is 455 kHz, an input signal at 455 kHz should be rejected.

To measure IF rejection, you must set the signal generator to the receiver's intermediate frequency and increase its output level until you achieve a 10 dB (S+N)/N ratio, then compare the generator's voltage output to the receiver's normal sensitivity, just as in the image rejection measurement. Typical IF rejection ratings are 40 to 60 dB. This means that a signal at the receiver's IF will be 40 to 60 dB below the desired signal.

DYNAMIC RANGE

Dynamic range is a measure of the receiver's ability to receive both very weak and extremely strong stations without introducing distortion to either. It is a function of the RF amplifier's linearity and the ability of the AGC system to control receiver gain.

To measure dynamic range, you first measure the sensitivity of the receiver for a 10 dB (S+N)/N ratio. This is the weak signal limit. You then increase the input RF signal until the output audio starts to distort. This marks the strong signal limit. The ratio between the weak signal limit and strong signal limit in dB is the dynamic range. Typcial values range for 100 to 120 dB and greater.

Unit 4

ANGLE MODULATION

INTRODUCTION

In the previous units, we discussed the different types of modulation, AM transmitters, and AM receivers. In this unit, we'll continue that discussion by covering angle modulation transmitters and receivers.

Recall that angle modulation refers to both phase and frequency modulation. In this unit, we will discuss both phase and frequency modulators, several types of demodulators, transmitters, receivers, and transceivers.

ANGLE MODULATION TRANSMITTERS

There are basically two types of angle modulation transmitters, depending on the modulation technique used. One type of transmitter uses a "direct" modulation process, where the master oscillator's frequency is varied in accordance with the modulating signal. The second type obtains an angle modulated wave "indirectly" by phase modulating the master oscillator's carrier signal in a later stage of the transmitter.

Reactance Modulator

The first type of direct modulation we'll discuss is called **reactance modulation**. Here, an FET or transistor stage is biased so that it presents a capacitive or inductive reactance to the master oscillator. When an audio signal is applied to the modulator, its reactance varies and, therefore, so does the master oscillator's frequency.

A typical reactance modulator is shown in Figure 4-1. In this circuit, C_1 and R_1 are connected directly across the master oscillator's tank circuit. The reactance of C_1 at the oscillator frequency is at least six times greater than the value of R_1. Therefore, the C_1 R_1 parallel branch appears almost purely capacitive. Thus, the voltage applied to the gate of Q_1 leads the oscillator voltage by 90° due to the action of C_1 R_1. This in turn causes the RF component of Q_1's drain current to also lead the oscillator voltage by 90°. Therefore, the entire circuit appears as a capacitive reactance across the master oscillator tank.

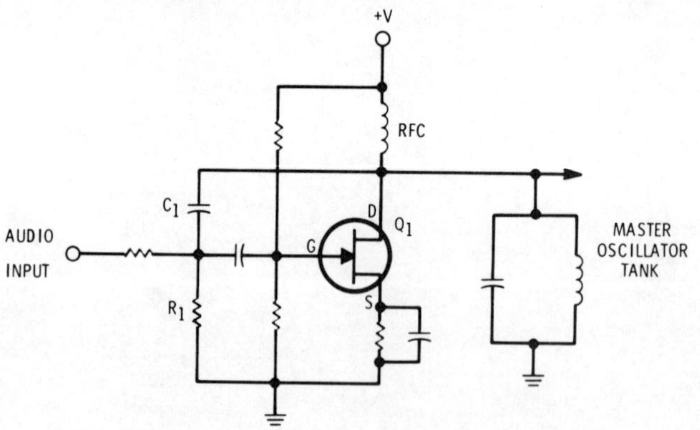

Figure 4-1
A reactance modulator.

Since the FET circuit acts as a capacitance, it controls the master oscillator frequency. When the audio input is zero, the capacitive reactance is at its normal value which is determined by the biasing components and C_1 R_1. When the audio input goes positive, the FET drain current increases and the capacitive reactance of the circuit also increases. This adds more capacitance to the oscillator tank and, as a result, the oscillator frequency is reduced. Likewise, when the audio input goes negative, the FET current decreases, also decreasing the capacitive reactance. This in turn increases the master oscillator's operating frequency.

A bipolar transistor could be used in place of the FET or even a vacuum tube. The circuit can also be biased to present a variable inductive reactance by reversing the C_1 R_1 connection. One disadvantage of the reactance modulator is that it cannot be used with a crystal controlled oscillator. Therefore, it requires some form of frequency control circuitry. This makes it an expensive and complex circuit, particularly when compared to the varactor modulator which is discussed next.

Varactor Modulator

A typical varactor diode modulator is shown in Figure 4-2. It consists of a varactor diode, D_1, in a Pierce crystal oscillator circuit. R_1 and R_2 reverse bias the varactor and set the diode's nominal capacitance value. This, in turn, establishes the oscillator's center frequency. When an audio signal is applied, it alternately adds to and subtracts from the varactor's reverse bias voltage. This changes the varactor's capacitance and, therefore, the oscillator's frequency.

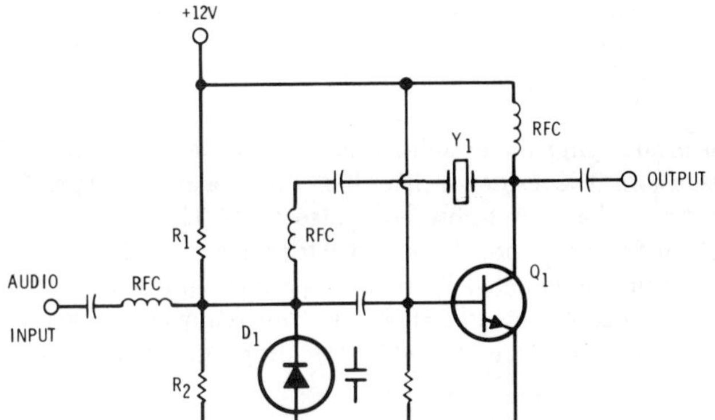

Figure 4-2
A varactor modulated crystal oscillator.

When the audio signal swings positive, the reverse bias on the varactor increases. This decreases the capacitance of the varactor which, in turn, causes the oscillator frequency to increase. When the audio signal swings negative, the reverse bias decreases. This increases the capacitance, forcing the oscillator frequency lower.

If the audio signal is a 1000 Hz tone, the oscillator frequency will swing above and below its center frequency 1000 times each second . Thus, the rate of deviation is the same as the audio signal. When the amplitude of the audio signal is increased, the varactor swings through a larger range of capacitance. Consequently, the amount of frequency deviation increases.

The varactor modulator is by far the most popular circuit used to obtain angle modulation. It offers simplicity, reliability, and the stability of a crystal oscillator.

Phase Modulator

Indirect angle modultion is obtained by using a phase modulator, such as that shown in Figure 4-3. The RF input signal is usually from the master oscillator or a buffer stage. This input RF is applied simultaneously to the gate of Q_1 and through C_1 to the output tank of L_1 and C_2. A phasor, or vector, diagram of this circuit is shown in Figure 4-4. Figure 4-4A shows the gate voltage phasor, E_g, and the tank voltage phasor, E_L. E_L is leading E_g by a given phase angle, Θ, due to the phase shift imparted by C_1. In addition, a second RF component appears across L_1 and C_2: the FET output drain voltage, E_D. This voltage leads E_g by 180° due to the FET's phase inversion. The phasor diagram of this is shown in Figure 4-4B. The output voltage of this circuit, E_O, is therefore the phasor sum of E_D and E_L as is shown in Figure 4-4B.

When an audio signal is applied to Q_1, the gate bias varies and so too does the stage gain. The result is a changing E_D phasor and in turn an E_O phasor that not only changes amplitude but also phase. This is shown in Figures 4-4B, C, and D. Therefore, the output of the phase modulator is both phase and amplitude modulated. To remove the amplitude variations, the next few stages must be operated at class C. This removes any AM, while the phase modulation is unaffected by the clipping action.

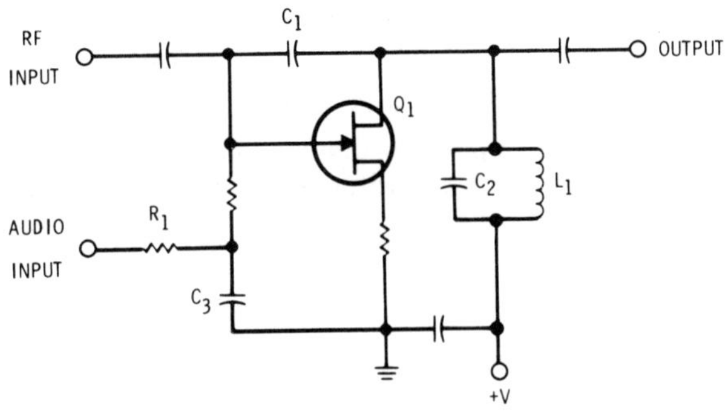

Figure 4-3
An FET phase modulator.

The advantage of the phase modulator is that the master oscillator is unaffected by modulation. Therefore, it offers increased frequency stability over the direct modulation techniques. However, it is difficult to obtain wide frequency deviations with a phase modulator. For this reason, its primary use is in narrow band, two-way FM communications rather than wide band FM broadcasting. You should also recall, from Unit 1, that a phase modulator can produce frequency modulation if the audio input passes through a low pass network before modulation takes place. This is accomplished in Figure 4-3 by the low pass RC network formed by R_1 and C_3.

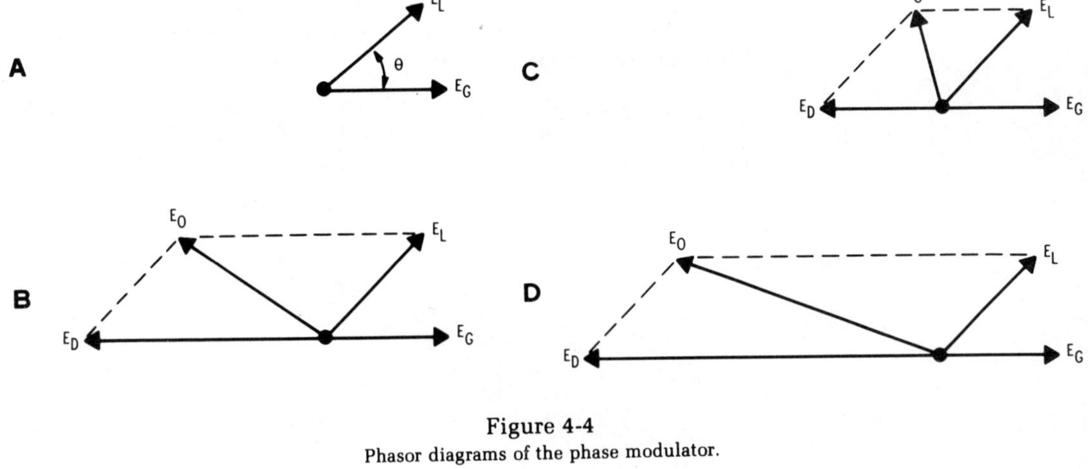

Figure 4-4
Phasor diagrams of the phase modulator.

The Complete Transmitter

The block diagram of an FM transmitter is shown in Figure 4-5. The audio signal is amplified and applied to the FM modulator. The modulator deviates the master oscillator according to the frequency and amplitude of the audio signal. The FM carrier is then amplified and transmitted.

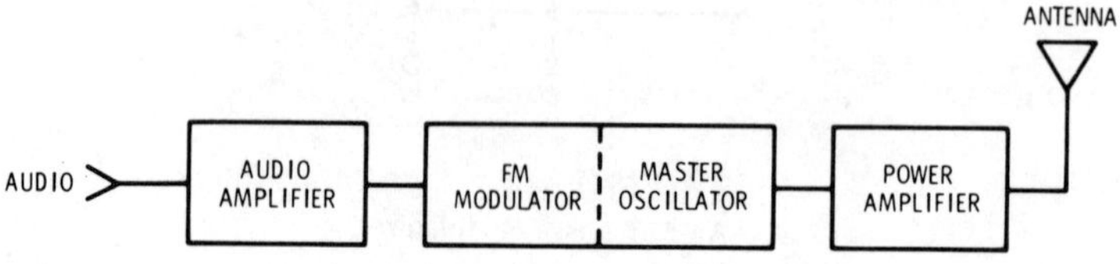

Figure 4-5
Simple FM transmitter.

This type of transmitter has several problems at high frequencies. For example, in an FM broadcast transmitter, the output signal will be between 88 and 108 MHz. This is a very high frequency and the master oscillator would have a great deal of difficulty generating a stable frequency carrier. Also, at this high frequency, it is difficult to shift the carrier the required amount without causing distortion.

Many of these problems can be solved by using a different approach. Figure 4-6 shows a block diagram of an FM transmitter that uses a much lower master oscillator frequency.

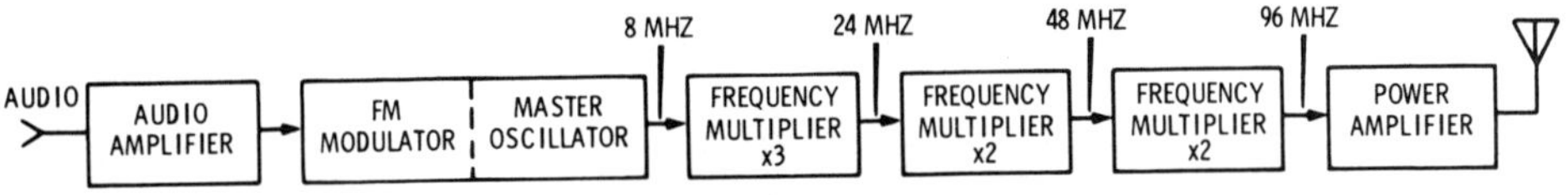

Figure 4-6
An FM broadcast transmitter.

Here the master oscillator frequency is only 8 MHz. To raise the frequency to the required 96 MHz, a series of frequency multipliers are used. Remember, a frequency multiplier is nothing more than a nonlinear amplifier whose output is tuned to a harmonic of the input frequency. The most common multipliers are doublers and triplers because the efficiency drops when higher multipliers are used. Here, one tripler and two doublers are used to multiply the 8 MHz signal by:

$$3 \times 2 \times 2 = 12$$

Thus, the output frequency is 8 MHz × 12 = 96 MHz.

The maximum deviation for FM broadcasting is ±75 kHz. Therefore, the output frequency must deviate this amount when an audio signal is applied. However, frequency multipliers increase carrier deviation as well as carrier frequency. For example, suppose the 8 MHz oscillator is varied from 7.9 MHz to 8.1 MHz. After this is multiplied 12 times, the carrier will vary from 94.8 MHz to 97.2 MHz. Thus, to obtain ±75 kHz deviation at the output, the master oscillator must be deviated:

$$\pm 75 \text{ kHz} \div 12 = \pm 6.25 \text{ kHz}$$

This can be easily accomplished with a varactor modulator.

One advantage that FM has over AM is that the power amplifier can operate in class C, which is more efficient. This is so because the FM waveform has a constant amplitude envelope. Therefore, only the frequency variations need to be reproduced accurately. Thus, a high efficiency class C power amplifier is used.

DEMODULATORS

There are many different ways to demodulate an angle modulated signal. The methods that are easiest to understand are not used very often, but to help you understand the more complex methods, we will discuss the simple ones first. We will then discuss the more complex (but more popular) circuits.

Slope Detector

The simplest FM detector is called a slope detector and is shown in Figure 4-9A. The two parallel resonant tank circuits L_1C_1 and L_2C_2, have an overall response curve like that shown in Figure 4-7B. Notice that the response curve does not peak at the center frequency of the IF signal. Instead, the center frequency falls half way up the side of the curve.

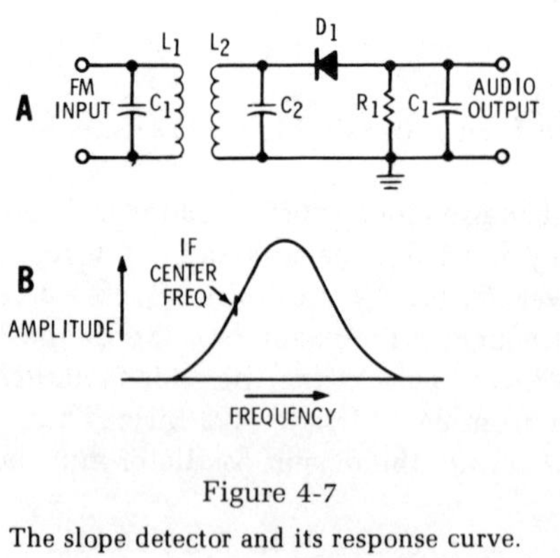

Figure 4-7

The slope detector and its response curve.

When the IF signal is at its center frequency, an average amplitude signal reaches D_1. When the IF swings above the center frequency, it approaches the peak of the response curve. Consequently, a higher amplitude signal is applied to the diode. When the IF swings below the center frequency, a low amplitude signal is applied to the diode.

In effect, the tuned circuit changes the frequency modulated IF signal to a signal which varies in amplitude. That is, it changes the FM signal to an AM signal. The diode then detects the AM signal just like the AM detector discussed earlier.

The disadvantage of this technique is that the response curve of the IF amplifiers which precede this stage must also be considered. Having the IF amplifier response curve offset is undesirable, because receiver gain is obviously reduced. Also, acceptable linearity is very difficult to achieve during the FM to AM conversion. Consequently, this technique is seldom used in practical applications.

Double-Tuned Detector

Another FM detector that is easy to understand is shown in Figure 4-8. This one is called a double-tuned detector. Transformer T_1 has a center-tapped secondary. The upper half of the secondary is tuned by C_1 to resonate slightly above the center frequency of the IF signal. Likewise, the lower half of the secondary is tuned by C_2 to resonate slightly below the center frequency.

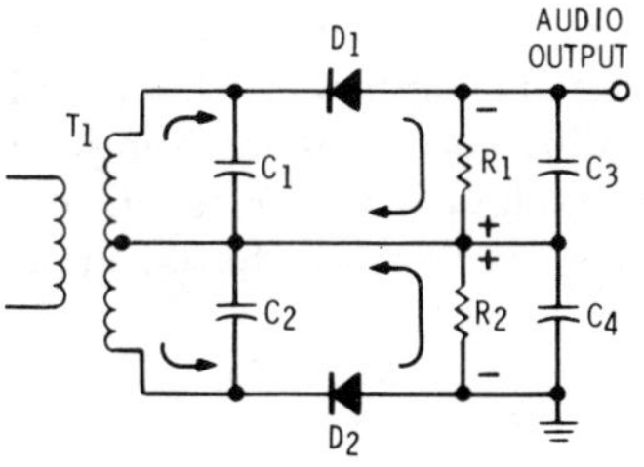

Figure 4-8
The double tuned detector.

When the IF signal is at its center frequency, both halves of the secondary are equally "detuned." D_1 and D_2 conduct equally as shown. Notice that the voltages developed across R_1 and R_2 are of opposite polarity and, therefore, tend to cancel. Consequently, if D_1 and D_2 conduct equally, the output voltage is zero.

Above the center frequency, the signal approaches the resonant frequency of the upper tank circuit. Thus, a larger signal is coupled to D_1 than to D_2; D_1 conducts harder developing a larger voltage across R_1 than is developed across R_2. Therefore, the resultant is a negative voltage.

When the carrier swings below the center frequency, the signal approaches the resonant frequency of the lower tank circuit. D_2 receives a stronger signal than D_1. D_2 conducts harder, developing a larger signal across R_2 than is developed across R_1. Since the voltage across R_2 is positive with respect to ground, the net output voltage is positive. As you can see, the output signal corresponds to the intelligence contained in the FM input signal.

The primary disadvantage of the double-tuned detector is the precise tuning required for each resonant circuit. If one of the tank circuits is slightly out of tune, the detected audio will be distorted. For this reason a double-tuned detector is seldom used to demodulate FM audio signals. However, it is used to demodulate frequency-shift-keyed or FSK signals. These are digital transmissions where a high level, or binary 1, corresponds to one frequency and a low level, or binary 0, corresponds to another frequency, which is slightly shifted from the first. Here, one resonant circuit is tuned to the binary 1 frequency and the other is tuned to the binary 0 frequency. The demodulated output is a pulse train which can be coupled to a digital computer or other peripheral device. Usually the FSK signal is at audio frequencies, for example 1070 Hz and 1270 Hz, and, therefore, the demodulator is also tuned to these frequencies. By using audio frequencies, the signal can be transmitted over the telephone lines.

Foster-Seeley Discriminator

A more popular FM detector is shown in Figure 4-9. This one is called the Foster-Seeley discriminator.

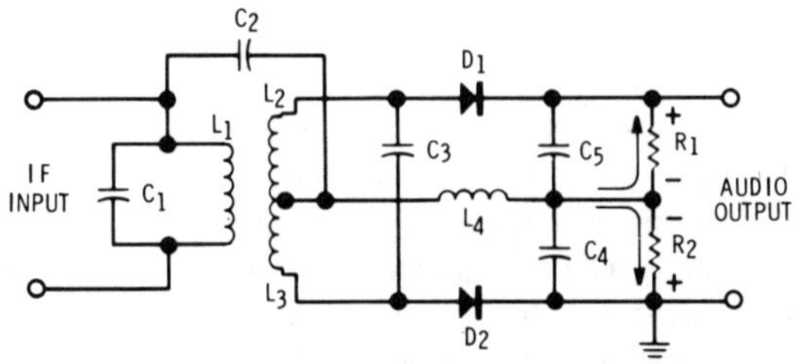

Figure 4-9
The Foster-Seeley discriminator.

The input to the circuit is an IF signal which is varying ±75 kHz at the audio rate. The output is the detected audio signal. The two diodes and their associated RC networks operate similar to their counterparts in the double-tuned detector discussed earlier. That is, when both diodes conduct, equal but opposite polarity voltages are developed across R_1 and R_2, the two voltages cancel, and the output is 0 volts. However, if D_1 conducts harder, the output is positive. By the same token, the output is negative when D_2 conducts harder. The audio signal can be recovered from the IF signal if:

1. Both diodes conduct equally at the center frequency;

2. D_1 conducts harder above the center frequency;

3. D_2 conducts harder below the center frequency.

What determines how much each diode conducts?

By transformer action, L_1 couples the IF signal to L_2 and L_3. As connected, L_2 and L_3 act as a center-tapped secondary. Thus, the voltage developed across L_2 (E_{L2}) is 180° out of phase with the voltage developed across L_3 (E_{L3}). E_{L2} controls the conduction of D_1 while E_{L3} controls the conduction of D_2. Keep in mind that these two voltages are equal in amplitude but are 180° out of phase.

The IF signal is also capacitive coupled through C_2 to L_4. A voltage which we will call E_{L4} is developed across L_4. E_{L4} controls the conduction of both D_1 and D_2.

The circuit is arranged so that E_{L4} is 90° out of phase with both E_{L2} and E_{L3}. However, as you will see, this is so only when the IF signal is at its center frequency. At the center frequency, E_{L4} leads E_{L3} by 90° but lags behind E_{L2} by 90°.

The amount that diode D_1 conducts is determined by E_{L2} and E_{L4}, while the amount that D_2 conducts is determined by E_{L3} and E_{L4}. We have seen that E_{L2} and E_{L3} are equal in amplitude but are 180° out of phase. At the center frequency, E_{L4} is 90° out of phase with both E_{L2} and E_{L3}. Thus, at the center frequency, E_{L4} adds to both of these signals equally. Consequently, D_1 and D_2 conduct equally and the output voltage is 0.

The parallel resonant circuit is resonant at the center frequency, where X_L exactly cancels X_C and the resonant circuit acts resistive. However, above resonance, X_L is larger than X_C. Thus, there is a net reactance that shifts the phase of E_{L2} and E_{L3}. E_{L2} is shifted more in phase with E_{L4}. Since E_{L3} is always 180° out of phase with E_{L2}, E_{L3} is shifted more out of phase with E_{L4}. That is, E_{L4} tends to add to E_{L2} but tends to subtract from E_{L3}. Thus, D_1 conducts harder than D_2. The net result is that the output swings positive each time the IF signal swings above the center frequency.

Below resonance X_C is larger than X_L. The net reactance shifts the phase of E_{L2} and E_{L3} in the opposite direction. This time, E_{L3} is shifted more in phase with E_{L4}. Consequently, D_2 conducts harder, producing a net negative output voltage. Thus, the output swings negative each time the IF signal swings below the center frequency.

If the modulating signal is a 1000 Hz tone, the IF will swing above and below the center frequency 1000 times each second. The discriminator produces an output sine wave that swings positive then negative at the same rate. Thus, the discriminator recovers the modulating signal.

One of the disadvantages of the Foster-Seeley discriminator, as well as the other FM detectors previously discussed, is that it is sensitive to amplitude changes. That is, any noise spikes present on the FM wave will be detected and passed on to the audio output as noise or "static." However, since an FM wave does not have an envelope that must be preserved, we can use a special circuit to clip off both the positive and negative peaks of the wave. The frequency variations will remain but the majority of noise impulses will be removed. This circuit is called a **limiter** and it immediately precedes the FM demodulator. The limiter is responsible for the noise-free reception of FM signals.

Ratio Detector

A further improvement over the Foster-Seeley discriminator is the ratio detector shown in Figure 4-10. This circuit is one of the most popular types of FM detectors because it has a significant advantage. It provides its own limiting action and requires no preceding limiter stage.

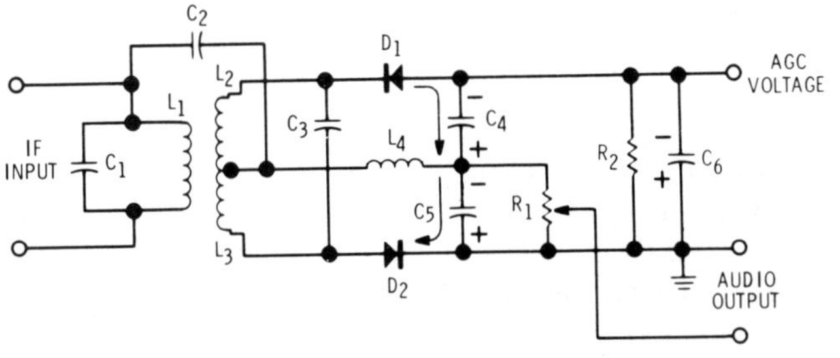

Figure 4-10
The ratio detector.

The ratio detector circuit looks somewhat similar to the Foster-Seeley discriminator. However, closer examination shows that diode D_1 is reversed and that the output configuration is different.

With D_1 reversed, the two diodes are in series across the entire secondary. Conduction of the two diodes is controlled by the same factors as in the Foster-Seeley discriminator. At the center frequency, the two diodes conduct equally. In this circuit, the voltages build up across C_4 and C_5 in series. Recall that in the discriminator, the diodes produced opposing voltages.

The key to the unique operation of the ratio detector is C_6. C_6 is a large value capacitor. After several cycles of the input signal, this capacitor charges to a voltage which is proportional to the average received signal strength. C_6 is large enough to hold the voltage across R_2 constant. It also holds the voltage constant across the series combination of C_4 and C_5. It does this even if there are momentary amplitude variations (noise). This is the reason that the ratio detector is relatively insensitive to noise.

C_6 also provides a convenient AGC voltage since the voltage across it is directly proportional to the input signal amplitude.

The voltage across C_5 plus the voltage across C_4 must always equal the voltage across C_6. At the center frequency, the two diodes conduct equally and the voltages across the two capacitors are equal. A sample of the voltage across C_5 is tapped from R_1. This is the audio output. This output will be at some negative DC level.

As in the discriminator, D_1 and D_2 alternately conduct harder as the IF signal swings above and below the center frequency. When the diodes are not conducting equally, the difference current flows through L_4. When D_1 conducts harder, the voltage across C_4 exceeds that across C_5. However, since the sum of these two voltages remains constant, the voltage across C_5 must decrease. Consequently, the voltage at the output must also decrease.

When D_2 conducts harder, a higher voltage develops across C_5. Therefore, the output voltage increases. As you can see, the output voltage swings in step with the frequency changes in the IF signal. That is, when the frequency increases, the output voltage increases. And, when the frequency decreases, the output voltage decreases.

The primary advantage of the ratio detector is its self-limiting action. It also provides AGC voltage.

Phase-Locked Loop Demodulator

It has always been possible to use phase-locked loops (PLL) as FM demodulators, but they were not practical until integrated circuits were developed. A complete phase-locked loop, like that shown in Figure 4-11, can be purchased as a single integrated circuit. This makes the PLL much simpler and cheaper to use. The basic loop of Figure 4-11 consists of a phase detector, low pass filter, and voltage controlled oscillator. The VCO or voltage controlled oscillator operates at the input frequency, in this case, the intermediate frequency. The phase detector compares the VCO frequency and the input frequency. It then develops an error voltage that is proportional to the amount and direction of the frequency difference. This error signal is then coupled to the low pass filter. The filter establishes many of the dynamic characteristics of the PLL. It determines the frequency range over which the loop will acquire and hold its phase lock. The filter also determines how fast the loop will respond to input frequency variations.

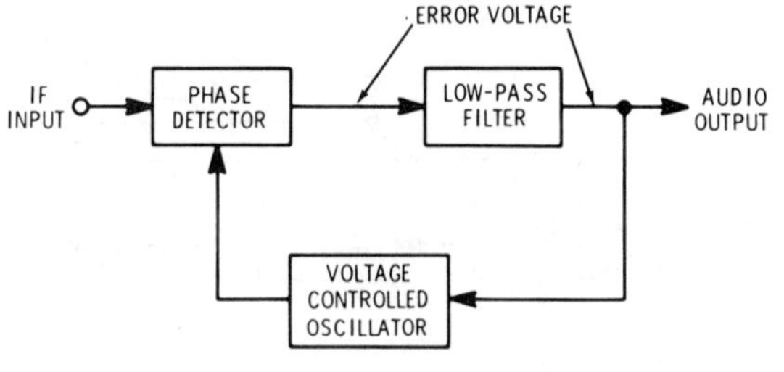

Figure 4-11
A phase-locked loop demodulator.

The error voltage from the low pass filter is then used to control the VCO. For example, if the input frequency deviates upward, an error voltage is generated by the phase detector. This voltage is filtered and applied to the VCO. The error voltage will cause the VCO to move upward in frequency in an exact lock with the input frequency. When the input signal is frequency modulated, the VCO tracks the FM deviation and, as a result, the error voltage is an exact duplicate of the modulating information.

Figure 4-12 shows a typical integrated circuit PLL. In this case, it is the Signetics NE565 monolithic IC. In this circuit, R_1 and C_1 are used to set the VCO's operating frequency range. C_2 and the internal 3.6 kΩ resistor form an RC low pass filter. The only difference between this circuit and the block diagram of Figure 4-11 is the DC amplifier which increases the error signal voltage.

The primary advantage of the phase-locked loop is its excellent performance at low cost, with a minimum of components.

It also eliminates costly inductors and transformers and it greatly simplifies tuning requirements.

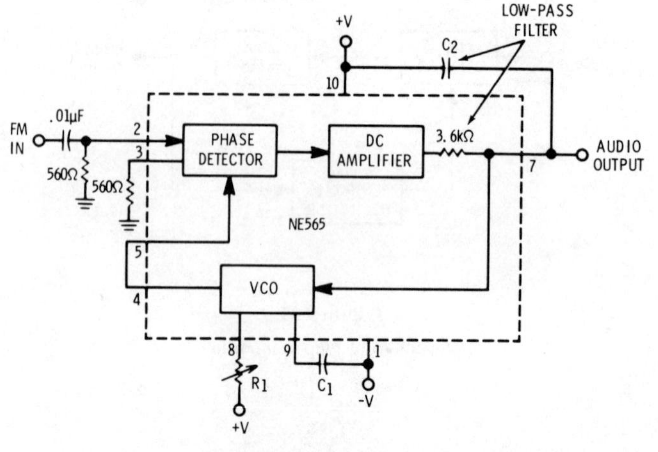

Figure 4-12

An IC phase-locked loop.

RECEIVERS

An FM receiver is actually very similar to the AM superheterodyne receiver discussed previously. That is, it converts the received RF signal to a lower intermediate frequency where the majority of gain and selectivity are obtained. The only difference is the use of an FM detector and possibly a limiter stage.

Block Diagram

A typical FM broadcast receiver is shown in Figure 4-13. Notice the similarity between this receiver and an AM receiver. The primary differences are the intermediate frequency, the limiter, and the FM detector. The IF of 10.7 MHz is chosen to provide good image rejection and also to provide the wider bandwidth required for FM broadcast signals. In FM broadcasting, maximum transmitter deviation is ±75 kHz. Therefore, the receiver bandwidth must be at least 150 kHz and is usually 180 to 200 kHz. The IF of 10.7 MHz allows this while maintaining acceptable IF gain.

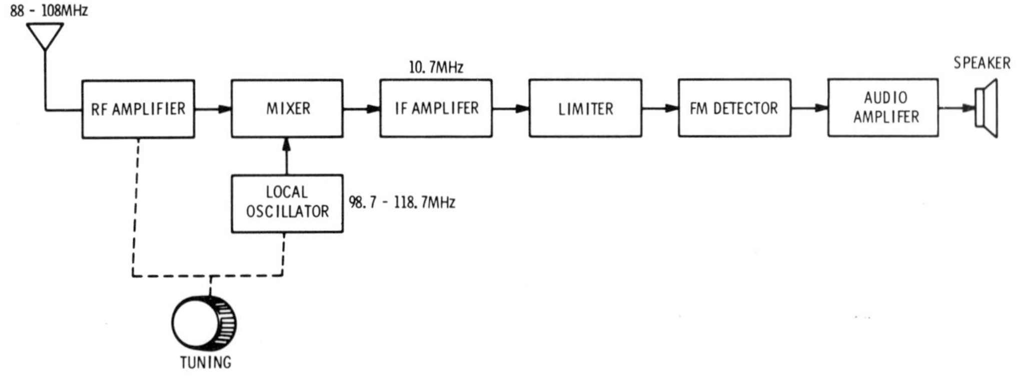

Figure 4-13
An FM broadcast receiver.

Let's follow the signal through the receiver. Assume that the receiver is tuned to 100 MHz. When the tuning indicator is set to this frequency, the RF amplifier is tuned to 100 MHz, while the local oscillator is tuned to 110.7 MHz. The RF and local oscillator signals heterodyne in the mixer, forming a difference signal of 110.7 − 100 = 10.7 MHz. This signal is passed to the IF amplifier where it receives most of its amplification. The IF amplifier also reduces the received bandwidth sufficiently to select the correct signal while rejecting all others.

If the received carrier is deviating ±50 kHz and the local oscillator frequency is constant, the 10.7 MHz IF signal will also deviate ±50 kHz. Thus, the original modulation is converted to the intermediate frequency without distortion.

LIMITER

The next stage in the block diagram is the limiter. As mentioned earlier, this circuit is used to remove any amplitude variations, such as noise, from the FM signal. It does this by saturating on the positive half cycle and going into cut off on the negative half cycle. It, therefore, clips or cuts off both the positive and negative peaks of the waveform and, along with them, most of the noise. This limiter action is shown in Figure 4-14. Note also that the frequency variations or original modulation is unaffected by the amplitude clipping.

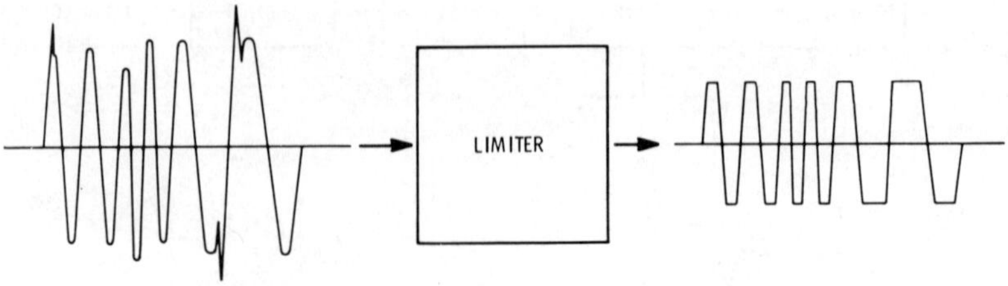

Figure 4-14
The limiter removes noise spikes and other amplitude variations.

Note from Figure 4-14 that the limiter stage must be overdriven to a certain extent. That is, if the input amplitude is not high enough to drive the limiter into cutoff and saturation, noise will pass on to the FM detector and eventually the audio output. In fact, one FM receiver specification refers to the "quieting" of the audio output. This is the sensitivity of an FM receiver. It specifies the unmodulated input signal required to reduce the audio noise output by 20 dB. The only way to reduce the noise output is to amplify the input signal sufficiently to overdrive the limiter. The limiter will clip the noise pulses from the wave and, in the absence of modulation, provide a "quiet" audio output.

AUTOMATIC FREQUENCY CONTROL

When an FM receiver is operated at VHF, as it is in FM broadcasting, the local oscillator must be extremely stable. For example, a frequency drift of just 0.1% at 100 MHz will cause the IF to drift 100 kHz. This will result in serious distortion of the demodulated wave. One way to prevent drift is to use a crystal controlled local oscillator. However, this is impractical for broadcast reception since you want to select many different frequencies without buying a new crystal each time. Although, crystals are used in two-way FM communications where only a few frequencies are required. In FM broadcasting, an automatic frequency control circuit is usually used to "lock" the local oscillator onto the incoming signal. Figure 4-15 shows the block diagram of an FM receiver that uses AFC.

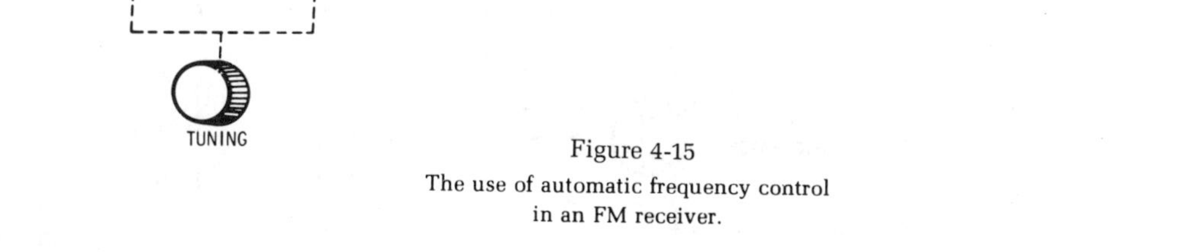

Figure 4-15
The use of automatic frequency control in an FM receiver.

The basic principle of AFC is to use a varactor, or a similar device, as part of the local oscillator tank. The varactor's capacitance is controlled by an "error" voltage that is produced when the oscillator goes off frequency. This error voltage is obtained from the output of the discriminator or ratio detector. Here, if the oscillator is on frequency, the output is zero. However, if the frequency drifts to one side or the other, an error voltage is produced. The polarity of this voltage indicates the direction of drift and the amplitude indicates the magnitude of frequency drift.

It seems that it would be a simple matter to use this error voltage to control the local oscillator. However, the output of the discriminator also contains the demodulated audio wave. Therefore, before this voltage can be used to control the oscillator, the audio component must be removed; otherwise, the AFC action would cancel the frequency modulation. The audio can be removed by using a low pass filter. This removes the audio while allowing the DC and the very low frequency error signal to pass on to the varactor.

Figure 4-16 shows a simplified AFC circuit. Q_1 and its associated circuitry form the local oscillator. C_2 is the receiver's main tuning control, while D_1 is a varactor which is in parallel with the oscillator tank. In this circuit, the output of the FM detector is coupled to both the audio amplifiers and the AFC circuit. R_1 and C_1 form a low pass filter which removes the audio component. Then this DC error voltage either adds to or subtracts from the varactor bias provided by R_2R_3. R_4 is used to isolate the AGC filter from the oscillator. As the varactor's capacitance changes, so too does the local oscillator frequency.

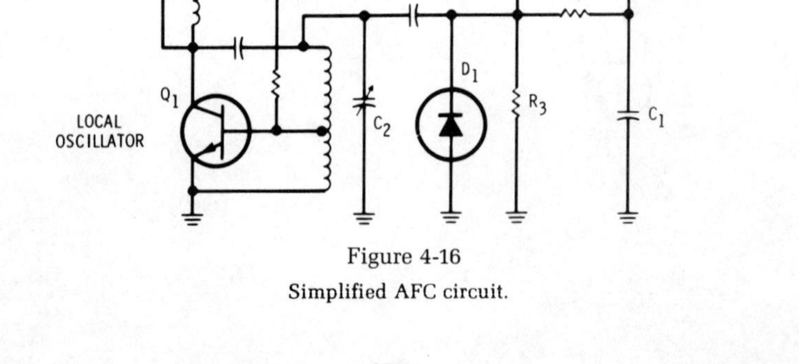

Figure 4-16
Simplified AFC circuit.

Careful observation of the AFC circuit of Figure 4-16 shows that it is actually a variation of the phase-locked loop. In this case, the FM detector is the phase detector, R_1C_1 are the filter, Q_1 and D_1 form the VCO, and the mixer is used to convert the two VHF signals (RF input and local oscillator) to a much lower frequency and to provide the needed 10.7 MHz offset.

Unit 5

PULSE MODULATION

INTRODUCTION

In amplitude and angle modulation, some characteristic of the carrier — amplitude, frequency, or phase — is **continuously** varied in accordance with the modulating information. However, in pulse modulation, a small sample is made of the modulating signal and then a pulse is transmitted. In this case, some characteristic of the pulse is varied in accordance with the sample of the modulating signal. The sample is actually a measure of the modulating signal at a specific time.

There are several types of pulse modulation systems. Three of the more common types are: pulse amplitude modulation (PAM), pulse duration modulation (PDM), and pulse position modulation (PPM). In each of these systems, a characteristic of the pulse — such as amplitude, duration, or position — is continuously varied in accordance with the modulating signal. This type of pulse modulation, where a pulse characteristic is continuously varied, is called **analog pulse modulation.**

Another type of pulse modulation is pulse code modulation (PCM), which is **digital pulse modulation**. With PCM, the modulating signal is sampled and then **quantized**. In quantization, each sample is assigned a specific numerical value according to its amplitude. This numerical value is then represented by a group of pulses of equal amplitude and duration. The absence or presence of pulses represent the modulating signal's value in the binary number system. This system has many advantages and, therefore, has many applications in modern communications.

You'll study both analog and digital modulation in this unit as well as multiplexing. This is a technique used to send many different communications channels over a common circuit.

ANALOG PULSE MODULATION

So far, we have divided pulse modulation into two broad categories: analog and digital. In analog pulse modulation, the indication of the sample amplitude is continuously variable, while in digital pulse modulation a numerical code is sent that indicates the sample amplitude.

In this section, you'll study analog pulse modulation in detail. You'll find that it can be subdivided into pulse amplitude modulation and pulse time modulation. Furthermore, you'll see that pulse time modulation includes pulse duration, pulse position, and pulse frequency modulation.

However, before you can study the various types of pulse modulation, we must discuss sampling and the sampling theorem.

Sampling

Sampling is just what the name implies. Rather than transmitting the entire modulating waveform, a periodic sample is taken and only the sample is transmitted. If enough samples are sent, the wave can be reconstructed at the receiving end. An example of sampling is shown in Figure 5-1. Here, a switching signal periodically energizes a relay and thereby connects the modulating signal to the output. If the sampling signal's frequency is high enough, sufficient information will appear at the output to accurately reproduce the modulating waveform. In fact, all that is needed to demodulate the output signal is a low-pass filter. This filter will remove the pulses and extract only the original information.

The critical part of this system is the sampling frequency. The minimum frequency is defined by the sampling theorem:

> *The sampling frequency in any pulse modulation system must be equal to, or exceed, twice the highest signal frequency in order to convey all the information of the original signal.*

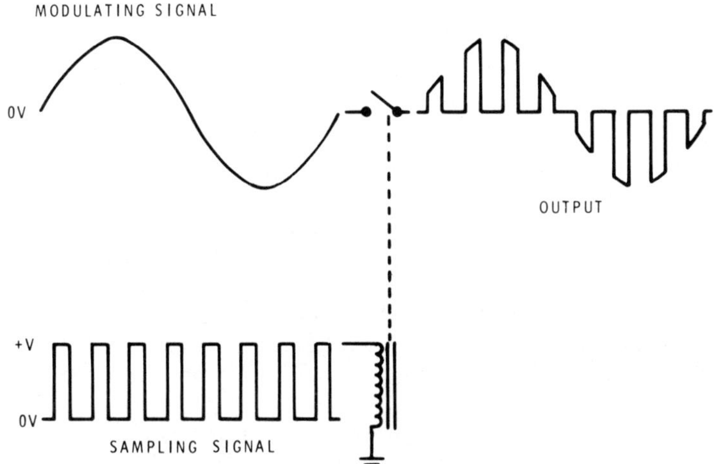

Figure 5-1
A simplified sampling system.

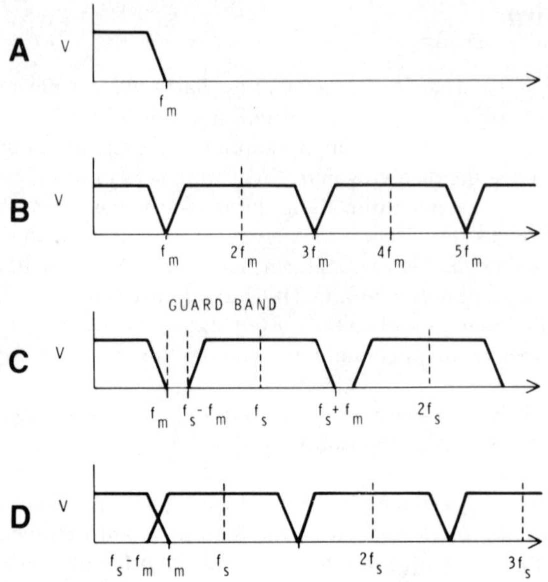

Figure 5-2
The effects of sampling frequency (f_s) on the signal's frequency spectrum.

The reasons for this are shown in Figure 5-2. Figure 5-2A shows the bandwidth of a modulating signal with a maximum frequency of f_m. This signal is then sampled at a rate of 2 f_m and the resulting frequency spectrum is shown in Figure 5-2B. At $2f_m$ there is a suppressed carrier with both upper and lower sidebands spaced around it. The lower sideband extends down to f_m since it is $2f_m - f_m = f_m$. The upper sideband extends to $2f_m + f_m = 3\ f_m$. Since the sampling signal is a pulse train, its harmonics theoretically extend to infinity as shown by carriers at 4 f_m, etc. In practice, the bandwidth is much less but still extends out quite a ways.

To demodulate the signal of Figure 5-2B, it is necessary to have an ideal low-pass filter with a cutoff frequency of f_m. If the filter's cutoff was infinitely sharp at f_m, it would pass the original modulating signal and nothing else.

Since ideal low-pass filters with infinitely sharp cutoffs are hard to come by, a sampling frequency of $2f_m$ is never used. Instead, a frequency slightly greater than $2f_m$ is used. The result is shown in Figure 5-2C. When the sampling frequency (f_s) is greater than $2f_m$, a guard band appears between f_m and the lower end of the first lower sideband. This guard band allows the use of a practical low-pass filter. However, the wider the guard band is, the greater the bandwidth of the pulse modulated signal becomes.

One example of a pulse modulation system is the telephone network. Here, the voice bandwidth is limited by filters to 3.3 kHz and the sampling frequency is 8 kHz. This is well above $2f_m$ and results in a guard band of $f_s - 2f_m$ or 8kHz − 6.6kHz = 1.4 kHz. This allows relatively simple low-pass filters to be used as demodulators. However, the guard band is not so wide as to unduly increase the signal bandwidth.

The situation shown in Figure 5-2D occurs when the sampling frequency is less than $2f_m$. Here, the overlap between the modulating signal and the lower sideband of f_s means that no amount of filtering will allow an exact recovery of the full modulating signal.

Figure 5-2 proves the sampling theorem since it shows that, in principle, the original signal can be recovered when $f_s = 2\ f_m$ (Figure 5-2B). It also shows why the minimum sampling frequency is $2\ f_m$ (Figure 5-2D). This minimum sampling frequency is called the **Nyquist rate.** Figure 5-2C proves that increasing f_s above the Nyquist rate provides a guard band which greatly eases the problems of filtering. However, increasing f_s also increases the signal bandwidth. Therefore, a compromise must be reached that allows sufficient guard band for filtering at a minimum signal bandwidth.

Pulse Amplitude Modulation

The simplest form of pulse modulation is pulse amplitude modulation (PAM). In PAM, the amplitude of the pulse varies in proportion to the amplitude of the signal. This is illustrated in Figure 5-3. The modulating signal is shown in Figure 5-3A and the sampling signal in Figure 5-3B. Figure 5-3C shows a dual-polarity PAM signal. This results if the waveform in "A" is centered on a zero-volt axis or, in other words, is a true AC wave. If a DC level is added to the modulating signal, single-polarity PAM results, as shown in Figure 5-3D. In this case, sufficient DC level is added to ensure that the pulses are always positive. Likewise, a negative DC voltage could be used to obtain negative pulses.

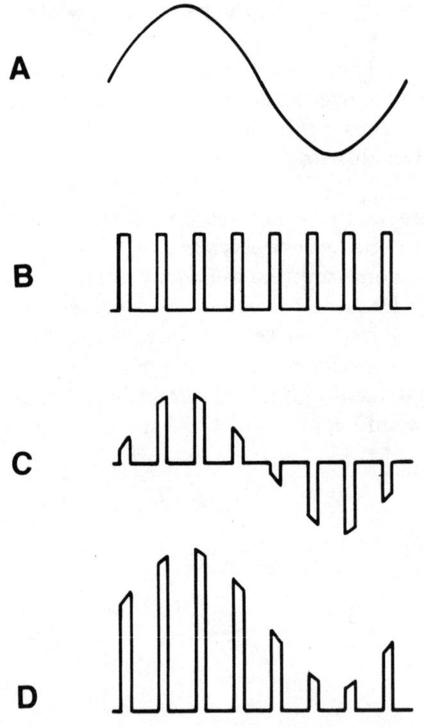

Figure 5-3
Examples of dual polarity and single polarity PAM.

One practical method of generating a PAM signal is shown in Figure 5-4. This circuit uses a 4016 integrated circuit CMOS switch. Basically, it is an FET logic switch. When the sampling pulse goes positive, the switch closes and the modulating input appears across R_3 and the output. When the sampling pulse drops to zero, the switch opens and the output is zero. As shown, the circuit provides dual-polarity PAM. However, single-polarity PAM can be achieved by adding R_1 and R_2. These resistors form a voltage divider that adds a DC level to the input signal. The result is that the input AC wave now varies around a positive DC reference rather than a zero-volt reference.

The demodulator for a PAM signal is merely a low-pass filter. It removes the sampling signal and its harmonics, and passes the original modulating signal. However, the rolloff of the filter must be steep enough to pass the highest modulating frequency and to fully attenuate the lowest sampling frequency component. That is, the filter's cutoff must fall well within the guard band of the particular PAM system.

When pulse amplitude modulation is used, it is sent over cable or wire, or it can be used to modulate a radio frequency carrier. When this is done, the PAM signal normally frequency modulates the carrier rather than amplitude modulate the carrier. However, PAM is not used very often to transmit information, since it is more susceptible to noise interference than other forms of pulse modulation.

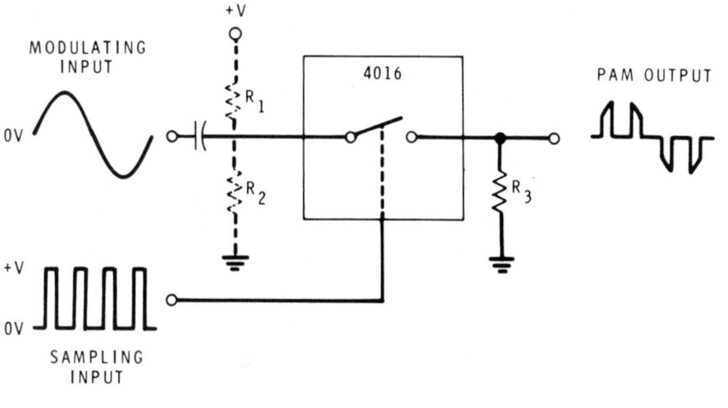

Figure 5-4
One method of obtaining PAM.

Pulse Time Modulation

In pulse time modulation (PTM), the modulating signal is sampled, just as it is in PAM. However, in PTM, the amplitude of the sample is indicated by a timing variation of the modulated pulse, rather than an amplitude variation. The variable timing characteristic may be the duration, position, or frequency of the pulses. Therefore, there are three basic types of PTM: pulse duration modulation, pulse position modulation, and pulse frequency modulation. This section will discuss the first two in detail. However, pulse frequency modulation has no significant practical applications and is, therefore, omitted.

In PTM, only the particular timing characteristic is modulated. The amplitude of each pulse remains constant. Therefore, amplitude limiters can be used at the receiver to remove most noise pulses or spikes. For this reason, PTM has the same advantage over PAM that frequency modulation has over amplitude modulation. Thus, PTM is used much more often than PAM in communications applications.

PULSE DURATION MODULATION

This type of PTM is also called pulse width or pulse length modulation, however, pulse duration modulation (PDM) is the preferred term. There are three different classifications of PDM: symmetrical PDM, leading edge PDM, and trailing edge PDM. These are shown in Figure 5-5 along with the sine wave modulating signal.

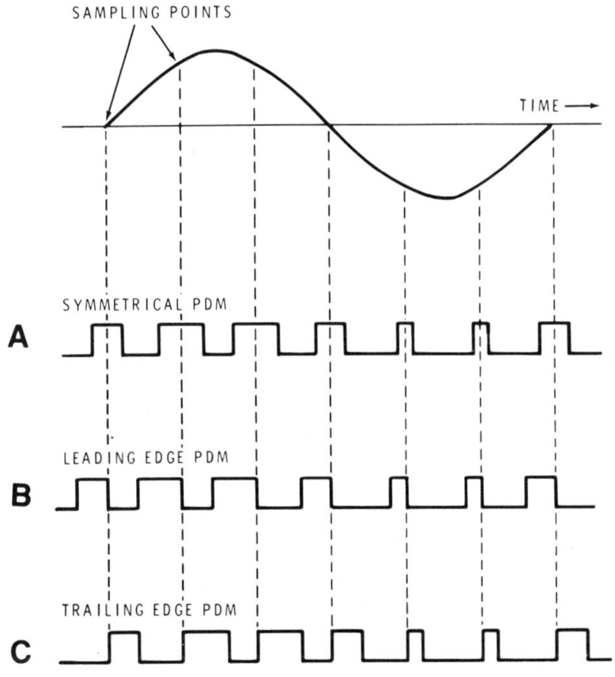

Figure 5-5
The three types of pulse duration modulation.

Figure 5-5A shows a symmetrical PDM waveform. Here, the modulating signal is sampled and both the leading and trailing edges of the pulse are varied in accordance with the sample amplitude. When the sample is a high positive voltage, the pulse duration increases, and when it is a high negative voltage, the pulse duration decreases. When the modulating wave is at zero, the pulse is at its average or reference duration. The spacing between the center of the pulses remains constant, as shown.

Leading edge PDM is shown in Figure 5-5B. In this type of PDM, the sample amplitude varies the leading edge of the pulse. The trailing edge of each pulse is fixed and, therefore, the spacing or timing between each pulse's trailing edge is constant.

Figure 5-5C shows trailing edge PDM. Here, the sample amplitude varies the trailing edge of the pulse, with the leading edge remaining fixed.

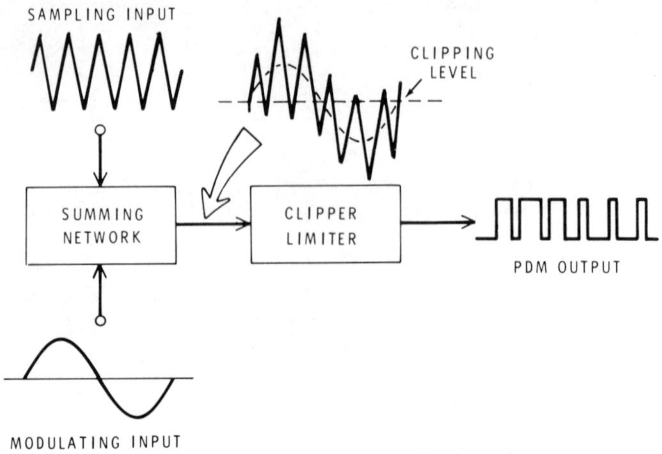

Figure 5-6
Method of generating a PDM signal.

A typical PDM generator is shown in Figure 5-6. There are two inputs: a triangular waveform at the sampling frequency, and the modulating input, which is a sine wave. These two signals are summed and the output is a triangle wave superimposed on the sine wave as shown. This wave is then applied to a clipper which removes everything below the clipping level shown. What remains is a triangle wave of varying amplitude and duration. This wave is then amplified and limited by several successive stages until all the pulses have the same amplitude, and both the leading and trailing edges are vertical. The result is a PDM wave. In this case, since the sampling wave is a triangle wave, the output is symmetrical PDM.

Leading edge PDM is generated using a sawtooth sampling wave that has a sloping leading edge and a vertical trailing edge. In this way, the resulting pulse has its leading edge modulated. Likewise, trailing edge PDM is generated using a sawtooth sampling wave with a vertical leading edge and a sloping trailing edge.

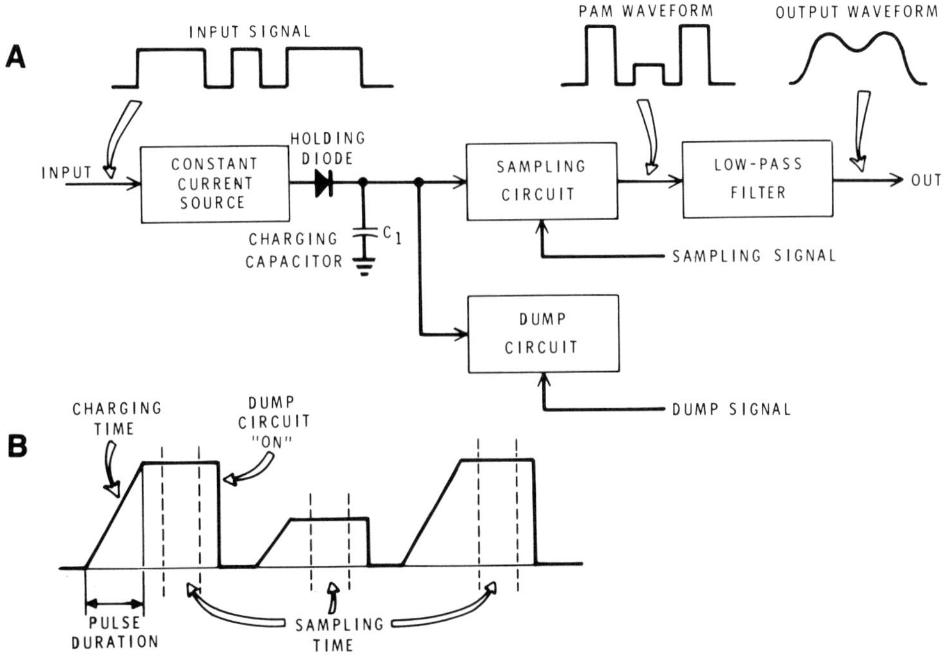

Figure 5-7
A PDM/PAM conversion demodulator.

The simplest method of demodulating a PDM wave is to use a low-pass filter. However, this technique can sometimes produce severe distortion. Therefore, a more elaborate but lower distortion system is usually used. One possible system is shown in Figure 5-7A. This circuit converts the PDM signal to PAM and then uses a low-pass filter to demodulate the PAM wave. This results in much lower distortion.

In Figure 5-7A the PDM input signal is applied to a constant current source. When the input goes positive, it turns on the constant current source, which then linearly charges C_1. C_1 continues to charge until the input signal drops back to zero. At this time, the current source is turned off and the voltage is held constant across C_1 by the holding diode which prevents any discharge. The sampling circuit is then turned on by a sample pulse which is derived from the input PDM signal. As soon as the sample is taken, the dump circuit discharges the capacitor and prepares it for the next pulse. The voltage waveform that appears across C_1 is shown in Figure 5-7B. The output of the sampling circuit is a PAM signal as shown. This is applied to the low-pass filter and the original modulating signal is recovered.

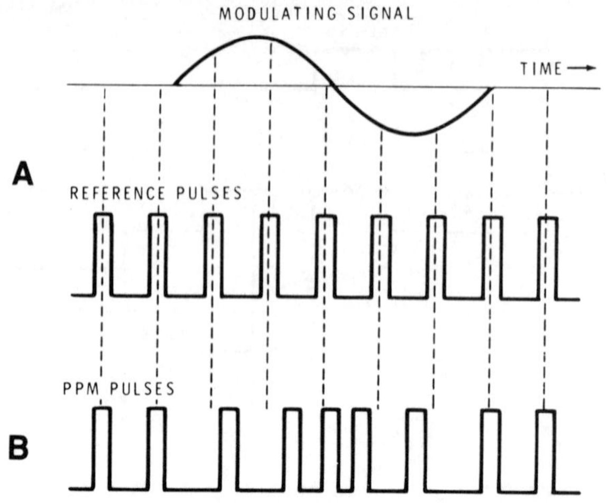

Figure 5-8
Pulse position modulation.

PULSE POSITION MODULATION

The next form of PTM is known as pulse position modulation (PPM). With this form of PTM, both pulse amplitude and duration remain constant while the position of the pulse, relative to a reference pulse, is varied in accordance with the modulating signal.

Figure 5-8B shows a typical PPM waveform. The modulating signal and the reference pulses are shown in Figure 5-8A. Note that when the modulating signal goes positive, the output pulse lags the reference pulse by an amount of time proportional to the sample amplitude. Similarly, when the modulating signal goes negative, the output pulse now leads the reference pulse by a proportional amount.

One method of generating a PPM signal is shown in Figure 5-9. This method uses the modulating signal to first generate a PDM waveform such as that shown in Figure 5-10. In this case, it is trailing edge PDM, where the leading edge is fixed and the trailing edge is modulated. The circuit of Figure 5-9 effectively "synchronizes" a fixed duration pulse to this modulated trailing edge. It does this by first differentiating the PDM signal, with the resultant waveform as shown in Figure 5-10. The positive spikes are then clipped off since they are fixed in time. The remaining negative spikes are synchronized to the trailing edge and are, therefore, position modulated. This clipped waveform is then used to trigger a monostable multivibrator which generates a fixed duration pulse for each input trigger. The duration of the pulse is determined by the monostable's RC time constant.

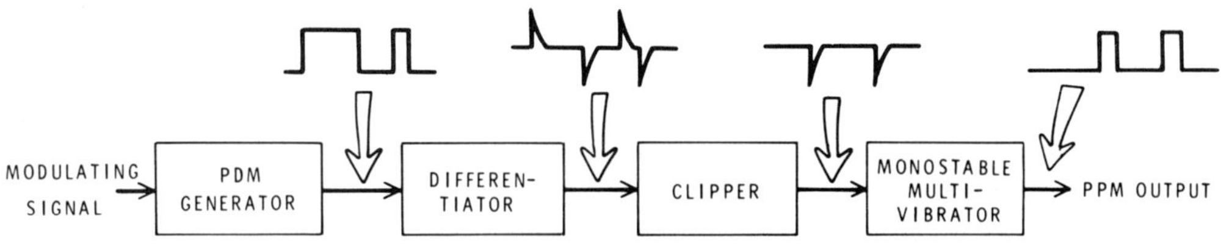

Figure 5-9
A pulse position modulation circuit.

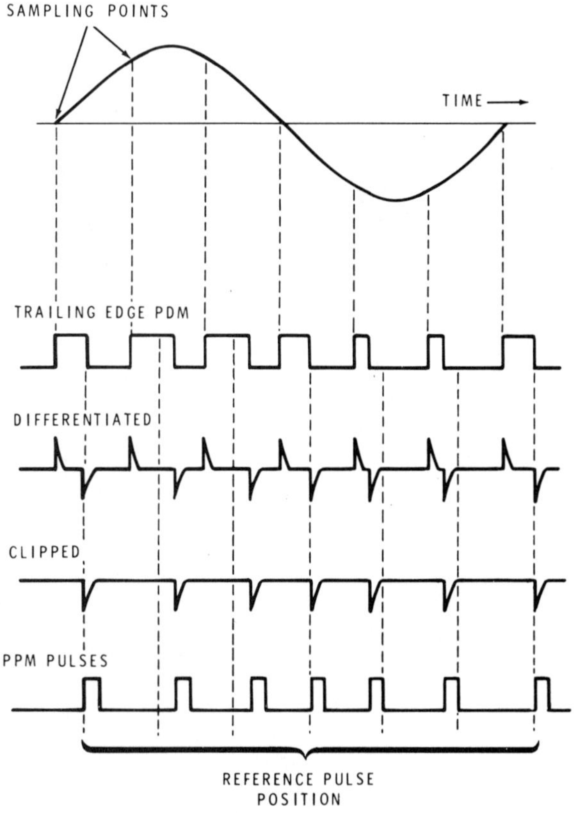

Figure 5-10
Waveforms for the PPM generator.

Figure 5-11 shows a typical PPM demodulator. The PPM signal is first applied to a limiter stage to remove any noise and to effectively regenerate the signal. It is then sent to a circuit that removes the synchronizing pulses sent by the transmitter. These pulses are used to lock the local reference pulse generator to the correct phase and frequency to demodulate the PPM signal. The reference pulses and the PPM signal are then applied to a flip-flop. Both these waveforms are shown in Figure 5-12. Note that the incoming PPM pulse lags the reference pulse by an amount proportional to its position modulation. As shown, in Figures 5-11 and 5-12, the reference pulse turns on the flip-flop and the PPM pulse turns it off. Therefore, the output of the flip-flop is a PDM wave whose pulse duration is directly proportional to the position of the input pulse. The resulting PDM signal is then converted to PAM as described earlier. The PAM wave is then coupled to a low-pass filter to recover the original modulating signal.

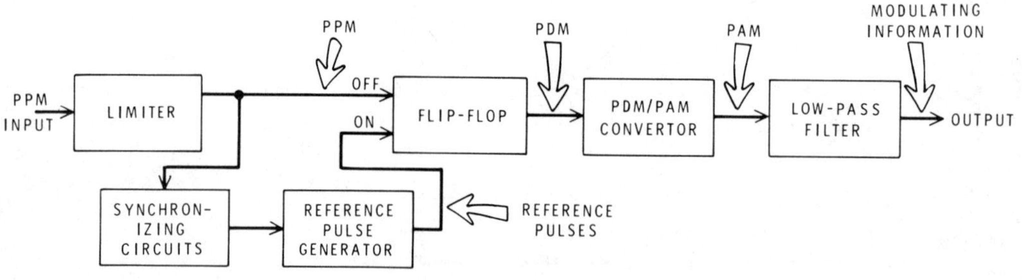

Figure 5-11
One method of demodulating a PPM signal.

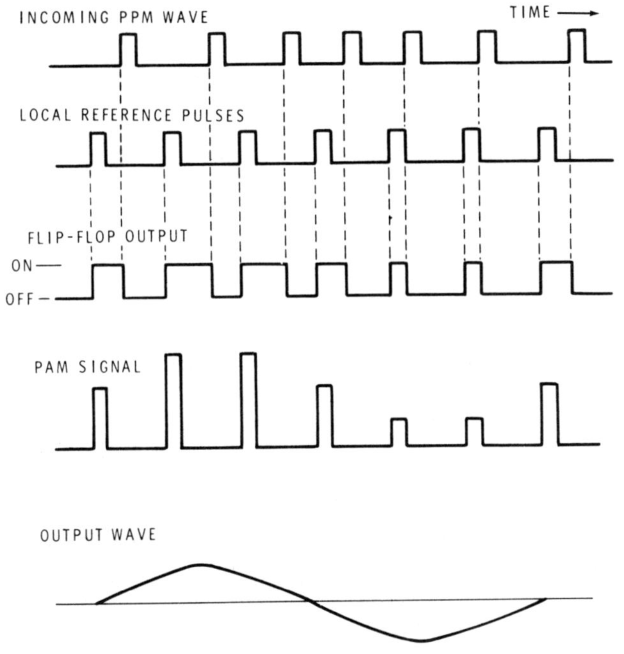

Figure 5-12
Waveforms for the PPM demodulator.

DIGITAL PULSE MODULATION

Pulse amplitude modulation and pulse time modulation are both susceptible to noise distortion. With PAM, it is quite obvious that once noise is present it is virtually impossible to remove it or to regenerate the signal. However, with PTM, a limiter can be used to remove most of the noise. Still, in PTM, the arrival of the leading and trailing edges of the pulses must be recorded precisely to accurately demodulate the signal. Since these pulse edges have finite rise and fall times, they too can be distorted to some degree by noise. For these reasons and others, digital pulse modulation (DPM) was developed. In this type of pulse modulation, groups of pulses are transmitted that represent binary numbers that correspond to the modulating voltage levels. Here, demodulation of the transmitted signal does not depend on the amplitude, duration, or position of the pulses, but only on their presence or absence. Under these conditions, it is relatively easy to recover pulses, even in the presence of high noise and interference levels. It is also a simple matter to regenerate this type of pulse modulation. As a result, digital pulse modulation is used almost exclusively in long distance communications.

Pulse Code Modulation

Pulse code modulation or PCM is the major form of digital pulse modulation. In PCM, the modulating signal is sampled, just as in other forms of pulse modulation. The sample amplitude is then converted into a binary code and transmitted as a stream of pulses.

In the other forms of pulse modulation, the sample amplitude is converted directly into pulse amplitude, duration, or position. However, in PCM, since the amplitude must be transmitted as a specific number out of a limited range of numbers, the sample amplitude must first be **quantized**. That is, each sample amplitude must be converted to the nearest standard amplitude or **quantum**. For example, suppose the PCM system has a total signal amplitude range of 7 volts and each 1-volt level corresponds to a specific binary code. Therefore, for this system each quantum or standard level is 1 volt. This is shown in Figure 5-13 for a sine wave signal with 8 quantum steps from 0V to 7V. Note that the quantizing waveform is actually a form of PAM, although it is limited to the quantum steps and is not continuously variable.

You'll notice that the first sampling point in Figure 5-13 is approximately 3.3V. Since there is no quantum level at this voltage, it is represented by the nearest level, which is 3V. This occurs at many places on the waveform. This error or distortion is called **quantizing noise**. It is noise because the errors are random. This is because the difference between the quantum level and the actual signal at any instant is completely unpredictable. The obvious method of reducing quantizing noise is to increase the number of quantum levels until the noise level is acceptable. However, increasing the number of levels increases the transmission bandwidth, so a compromise must be made between acceptable quantizing noise and bandwidth.

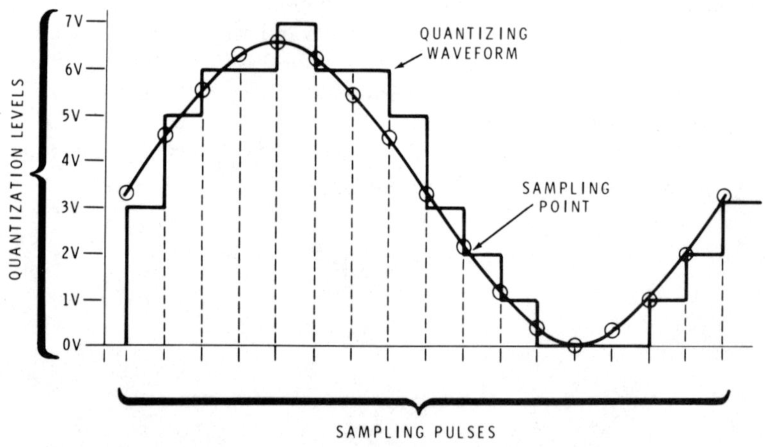

Figure 5-13
A quantized sine wave.

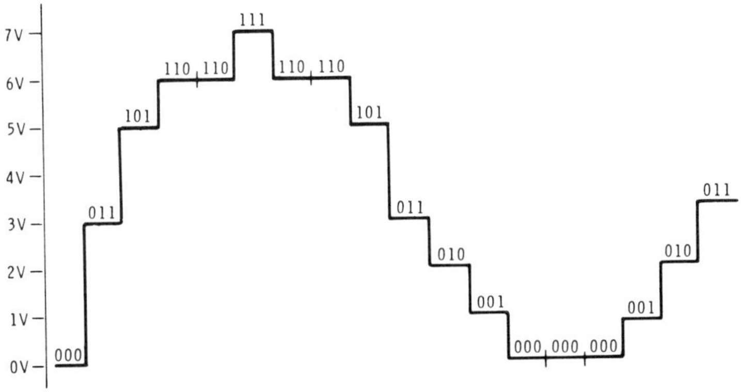

Figure 5-14
Coding the quantized wave.

After quantization has occurred, each sample must be coded as a binary number before it can be transmitted as PCM. Figure 5-14 shows the results of coding the quantized waveform from Figure 5-13. Since there are only 8 quantum levels, they can be represented by a 3-bit binary word, with 000_2 representing 0V and 111_2 representing 7 V. Once the quantizing waveform is coded, each sequential sample is transmitted as a pulse code. A table comparing the quantizing level, binary number, and pulse code is shown in Figure 5-15.

QUANTIZING LEVEL	BINARY NUMBER	PULSE CODE
0V	000	
1V	001	
2V	010	
3V	011	
4V	100	
5V	101	
6V	110	
7V	111	

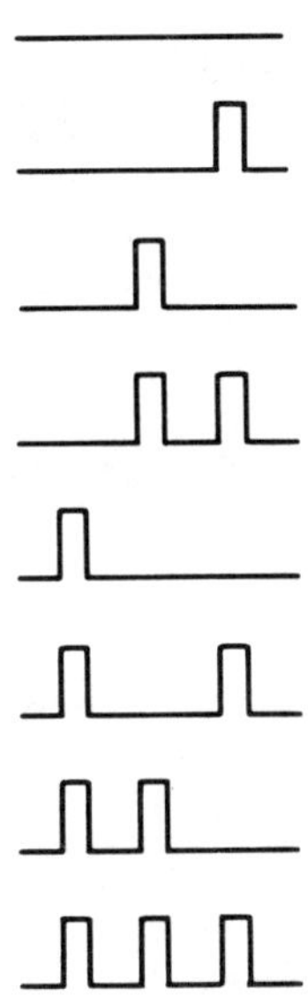

Figure 5-15
Three bit PCM.

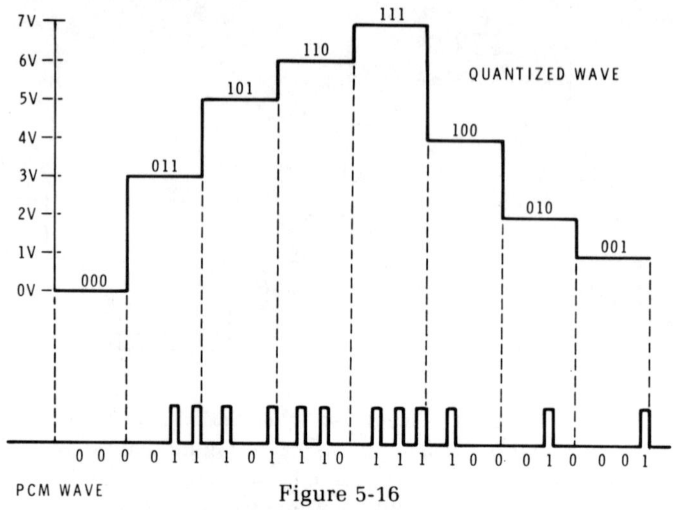

Figure 5-16
A PCM waveform.

Figure 5-16 shows the resultant PCM wave for a typical quantized wave. In practice, synchronizing pulses are required to insure that the receiver decodes the correct pulses. Also, in practical PCM systems, an 8-bit binary word is often used instead of the 3-bit word shown here. The 8-bit word provides 256 quantizing steps, which allows much better reproduction of the modulating signal with very little quantizing noise.

A simplified diagram of a PCM transmitter is shown in Figure 5-17. It consists of two sections. The first section is an analog-to-PAM converter that performs the quantizing process. The second section is a PAM-to-PCM converter that does the binary coding. The total system is actually an analog-to-digital converter.

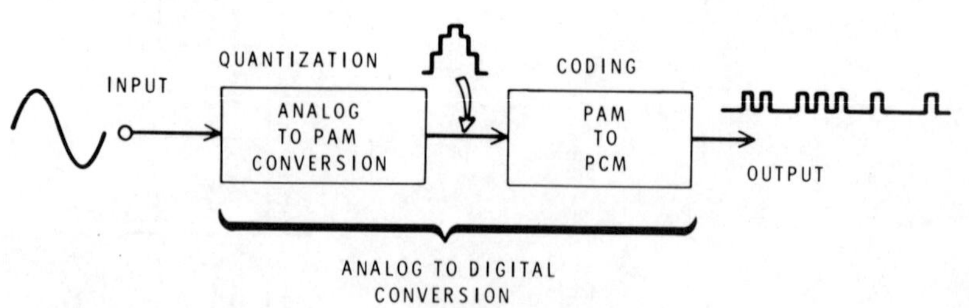

Figure 5-17
A PCM transmitter.

A PCM receiver is shown in Figure 5-18. It is made up of a PCM-to-PAM converter and a low-pass filter to convert that PAM back to the original modulating signal. It is, in essence, a digital-to-analog converter.

As mentioned earlier, the primary advantage of PCM is its much better immunity to noise and interference. For example, a typical PCM transmission can be sent over a communications channel having a signal-to-noise ratio of 21 dB with minimal error. In fact, the error would be just one pulse missed or decoded improperly every 17 minutes. If the signal-to-noise ratio is improved to 23 dB, the error rate drops to one error every four months! To achieve this low error rate in an AM system would require a signal-to-noise ratio of 60 to 70 dB.

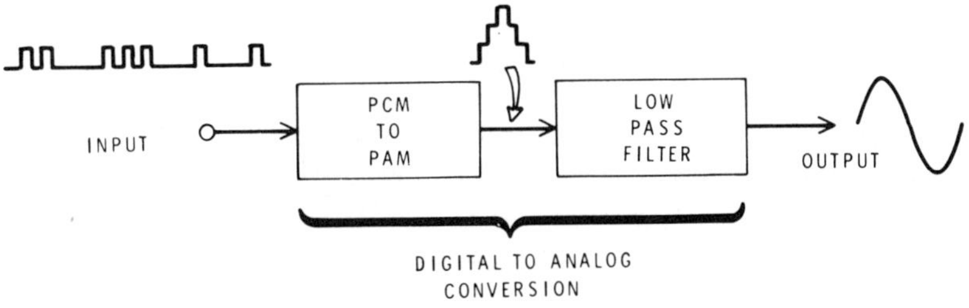

Figure 5-18
A PCM receiver.

Other DPM Systems

There are many other digital pulse modulation systems that are variations of PCM or slightly different systems. Two of the more common types are **differential PCM** and **delta modulation**.

Differential PCM is very similar to ordinary PCM. However, in this system, each binary word indicates the difference in amplitude between the present sample and the previous one. It specifies whether the change is positive or negative and the magnitude of the change. Therefore, the PCM code needs to indicate only the relative amplitude of the sample rather than the absolute amplitude as in normal PCM. This system takes advantage of the redundancy of most modulating signals, such as speech. That is, each sample is related to the previous sample; large variations from one to the next are unlikely. For this reason, it takes fewer bits to indicate the relative change in amplitude than the absolute amplitude. Thus, differential PCM has a smaller bandwidth than normal PCM. However, complications in encoding and decoding presently outweigh any bandwidth advantage. As a result, differential PCM has few applications.

Delta modulation is basically a 1-bit differential PCM system. In the simplest form of delta modulation, just one bit is sent per sample and its polarity indicates whether the signal is larger or smaller than the previous sample. An example of this is shown in Figure 5-19. Figure 5-19A shows the quantized wave and Figure 5-19B shows the delta modulation pulse train.

This system has several advantages: greatly simplified encoding, decoding and quantization. However, it cannot handle rapid amplitude variations between samples unless the sampling frequency is very high. Therefore, delta modulation requires a higher sampling rate and larger bandwidth than a comparable fidelity PCM system. Its advantages somewhat offset this disadvantage and, as a result, its use is increasing.

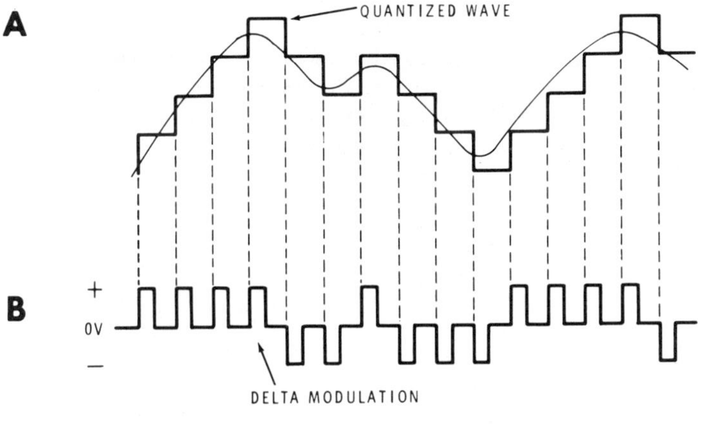

Figure 5-19
Delta modulation.

MULTIPLEXING

Multiplexing is the process of transmitting several separate information channels over the same communications circuit simultaneously without interference. There are two basic types of multiplexing: time division multiplex (TDM) and frequency division multiplex (FDM).

Time Division Multiplex

In TDM, several information channels are transmitted over the same communications circuit simultaneously using a time sharing technique. As an example, PAM waveforms can be generated that have a very low duty cycle. This means that if a single channel is transmitted, most of the transmission time would be wasted. Instead, this time is fully utilized by transmitting pulses from other PAM signals during the intervals. A PAM-TDM waveform for three channels is shown in Figure 5-20. The first pulse is a synchronizing pulse which is used at the receiver in demultiplexing. The second pulse is amplitude modulated by channel 1, the third by channel 2, and the fourth by channel 3. This set of pulses is called a **frame**. Four complete frames are shown in Figure 5-20.

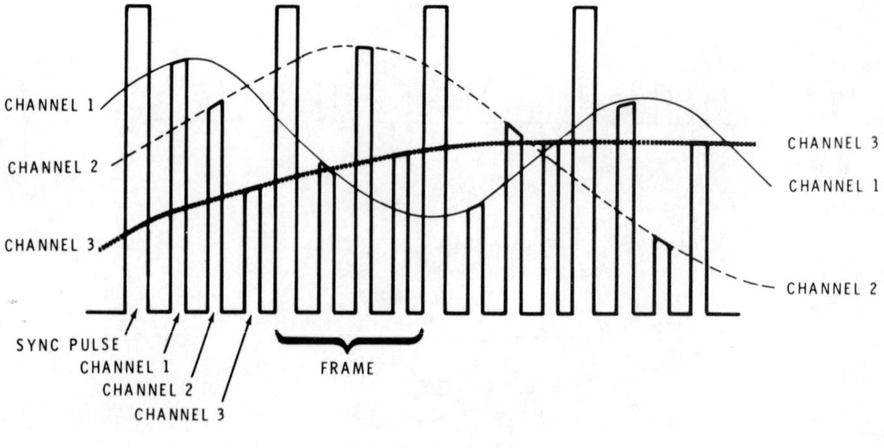

Figure 5-20
Three channel time division multiplex using single polarity PAM.

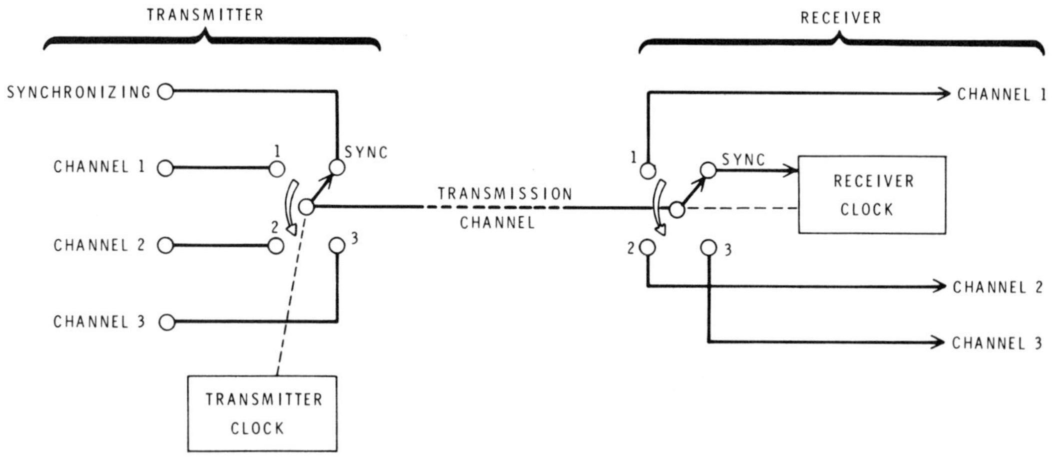

Figure 5-21
A very simplified diagram of a TDM transmitter and receiver.

A simplified diagram of a TDM transmitter and receiver is shown in Figure 5-21. In the transmitter, a rotating switch, called a commutator, connects each channel to the output in sequence. The speed of the commutator is set by the transmitter clock. Note that a synchronizing channel is also sent. This locks the receiver clock in phase with the transmitter clock.

Once the TDM signal reaches the receiver, it is demultiplexed, or decommutated, by another rotating switch. This switch routes the pulses to the correct channel. It also routes the sync pulse to the receiver clock to keep the transmitter and receiver synchronized. In practical TDM transmitters and receivers, the rotating switches are replaced by electronic switches.

The primary advantage of TDM is that several channels of information can be transmitted simultaneously over a single cable, a single radio transmitter, or any other communications circuit. Also, any type of pulse modulation may be used in TDM. In fact, many telephone systems use PCM - TDM.

Frequency Division Multiplex

Frequency division multiplex or FDM is normally an analog technique and, therefore, doesn't use pulse modulation. However, it is included here to complete the topic of multiplexing.

FDM, like TDM, is used to transmit several information channels over the same communications circuit simultaneously. However, in FDM, each channel uses a different band of frequencies to avoid interference. Thus, the distinction between FDM and TDM becomes apparent. In TDM, each channel occupies the entire frequency spectrum for only a fraction of the time. In FDM, each channel *continuously* occupies a small fraction of the transmission spectrum.

Let's examine a simplified FDM system. Figure 5-22 shows the block diagram of a 5-channel FDM transmitter. Each channel is a telephone voice circuit whose bandwidth is limited to 3.5 kHz. Channel 1 is transmitted just as it is, so it occupies the spectrum from approximately 100 Hz to 3500 Hz. Next, a 4 kHz pilot carrier is transmitted. It is used at the receiver in the demultiplexing of the FDM signal.

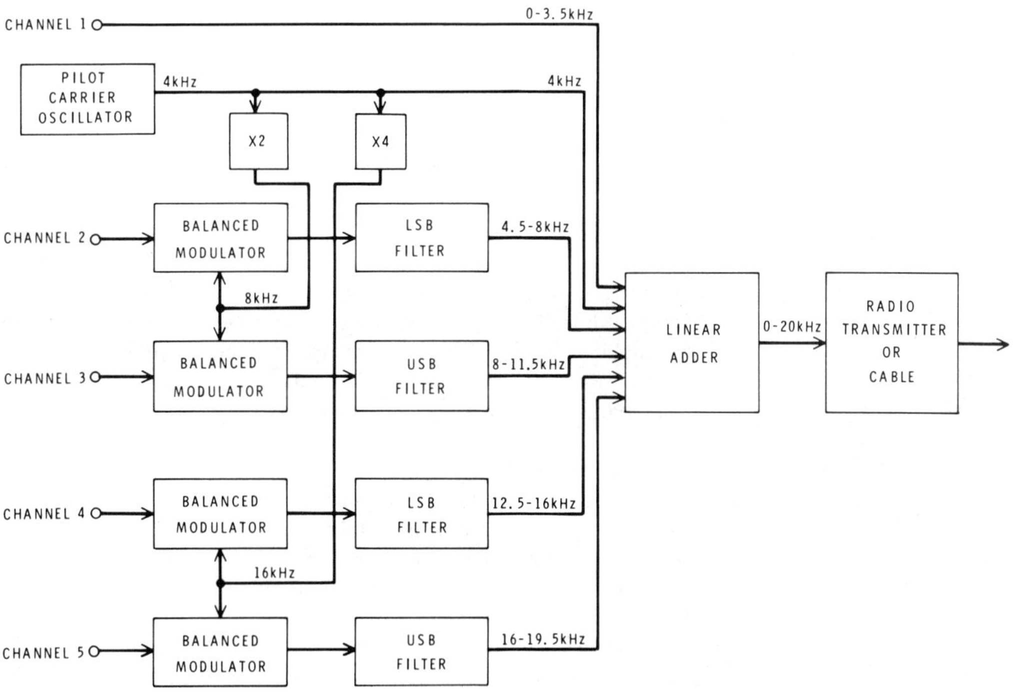

Figure 5-22
An FDM transmitter.

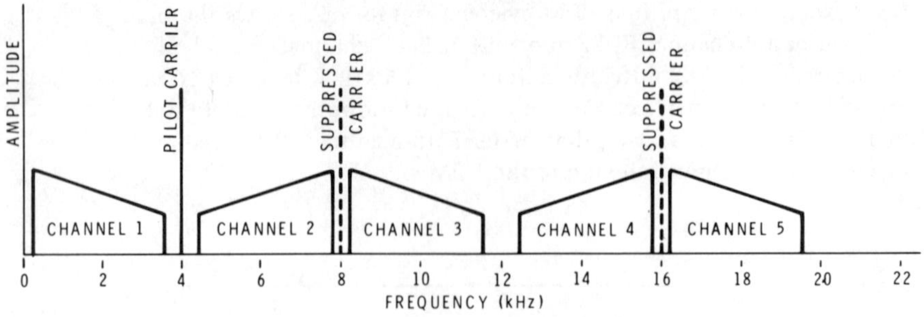

Figure 5-23
The output spectum of the FDM transmitter.

Channel 2 is applied to a balanced modulator along with an 8 kHz subcarrier, which is derived from the pilot carrier. The output of the balanced modulator is applied to a lower sideband filter. The result is a lower sideband, suppressed-carrier signal that extends from 4.5 kHz to 8 kHz. Channel 3 is also applied to a balanced modulator along with the 8 kHz subcarrier. However, it becomes an upper sideband signal extending from 8 kHz to 11.5 kHz. Likewise, channels 4 and 5 become lower and upper sidebands, respectively, on a 16 kHz subcarrier.

The output frequency spectrum of this transmitter is shown in Figure 5-23. Here, channel 1 occupies the spectrum from 100 Hz to 3.5 kHz. It is shown as having a lower amplitude at the higher frequencies to represent average speech content. The pilot carrier is at 4 kHz and channel 2 is from 4.5 kHz to approximately 8 kHz. Note that channel 2's frequency spectrum is inverted. This is because it is a lower sideband signal. The same is true for channel 4. On the other hand, channels 3 and 5 are not inverted since they are upper sideband signals. This composite FDM signal is then used to modulate a radio transmitter, which may use any form of modulation, or it is sent on a telephone line or cable.

The block diagram of a 5-channel FDM receiver is shown in Figure 5-24. Here, each specific frequency band is separated by filters. This demodulates channel 1 immediately. The 4 kHz pilot carrier is also separated and used to synchronize an oscillator. It, in turn, provides the necessary subcarriers for demodulation. Channels 2 through 5 are each separated and applied to a balanced modulator along with the appropriate subcarrier. The audio channels are then separated from the balanced modulators' outputs by low-pass filters. Thus, each channel has been recovered.

One disadvantage of FDM is that the transmission amplifiers and other circuits must be extremely linear. This is because any nonlinearity will cause harmonic distortion and cross modulation. For example, any harmonics from channel 1 will appear in all the other channels, to a certain extent. Also, if channels 2 and 3 mix, their cross modulation products will appear in channels 4 and 5 as well as channel 1. This can cause severe distortion. With a TDM signal, this distortion still occurs, but since the signal occupies the entire spectrum at one time, it does not interfere with the other channels.

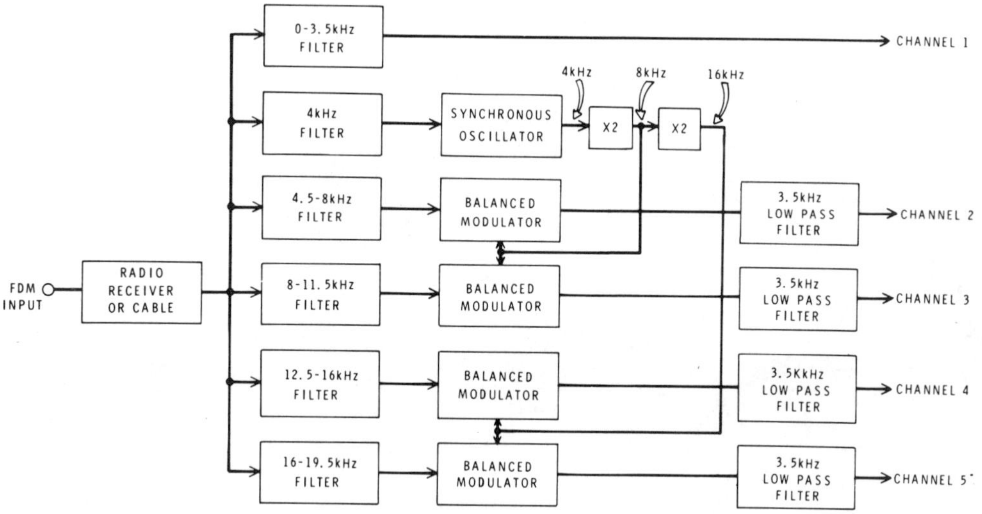

Figure 5-24
A 5-channel FDM receiver.

Unit 6

ANTENNAS

INTRODUCTION

An antenna, usually a wire or collection of wires, converts high frequency current into electromagnetic waves for transmission and vice-versa for reception. In previous units on transmitters and receivers, antennas were not discussed but were assumed to exist at the output of transmitters and at the input of receivers. This unit discusses antennas and antenna systems.

There are three main elements in the antenna "system." These are the antenna itself, the transmission line, and the transmitted radio wave's path. The transmission line transfers the RF energy to the antenna which then radiates an RF wave. The atmosphere then influences, to a certain degree, the path that the radio wave follows. This unit discusses all three of these important items in detail.

TRANSMISSION LINES

The primary purpose of a transmission line is to transfer energy from a source to the load. A good example is the transmission line used to transfer RF energy from a transmitter to the antenna. Lines are also used to interconnect equipment such as transmitters to power amplifiers, receivers to speakers, etc. When the energy being transferred is DC or audio, the lines present no special problems. At these frequencies, they usually appear as short circuits and, for most purposes, can be ignored. However, at higher frequencies, transmission lines take on very peculiar characteristics that cannot be ignored. These characteristics are primarily due to the signal's wavelength, the distance the wave travels in one cycle. This section discusses these characteristics in detail.

The Balanced Transmission Line

Most of us are familiar with the "TV twin lead" type of transmission line shown in Figure 6-1A. This is the type of lead-in wire used to connect an antenna to a television receiver. This type of transmission line is known as a **balanced** line. Both wires of the line carry RF current and the current in each wire is 180° out of phase with the other wire. (Shown in Figure 6-1B). Therefore, the current in this type of transmission line is balanced with respect to electrical ground. On the other hand, with an **unbalanced** transmission line, one wire would be at ground potential, while the other wire would carry the RF current.

A

B

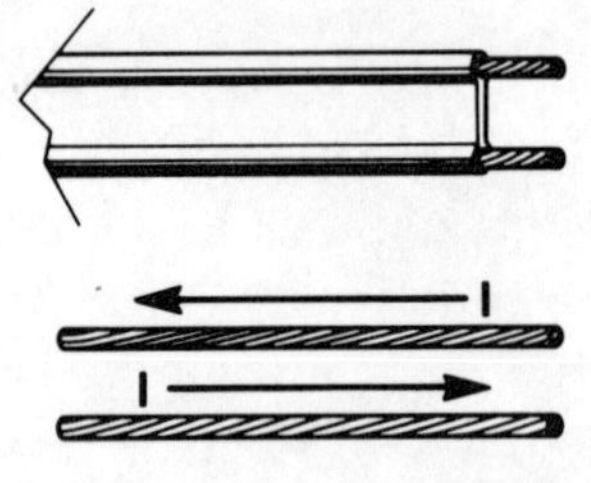

Figure 6-1
Balanced transmission line.

The equivalent circuit of a balanced transmission line is shown in Figure 6-2. Inductance is present because, whenever current flows through a conductor, a magnetic field is produced around it. Since a potential difference exists between the two conductors of the transmission line, due to the 180° phase difference, an electrostatic field exists between the conductors. Therefore, some small amount of capacitance also exists. The equivalent circuit shown in Figure 6-2 represents the ideal transmission line. In a practical line, a series resistive element is present due to the resistance of the conductors. Also, a parallel resistive element is present due to leakage in the dielectric material that separates the two conductors. The inductive, capacitive, and resistive properties of the transmission line present a specific impedance to any AC waveform.

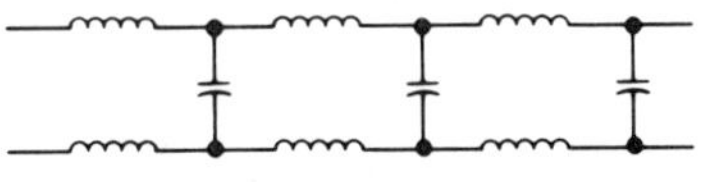

Figure 6-2
Balanced line equivalent circuit.

If an RF generator is connected to an infinitely long transmission line, a forward traveling wave of RF energy will start at the generator and travel, at the speed of light, down the line. This is shown in Figure 6-3A. This applied RF energy causes a specific amount of current to flow in the transmission line and a resulting potential difference or voltage to appear across the two conductors. If you now take current and voltage measurements, with an AC meter, at several points on this line, you will find that the amplitude of voltage and current is constant at all points. That is, the applied RF energy causes a specific current and voltage to appear at all points on the transmission line. A graph of voltage and current on this transmission line is shown in Figure 6-3B.

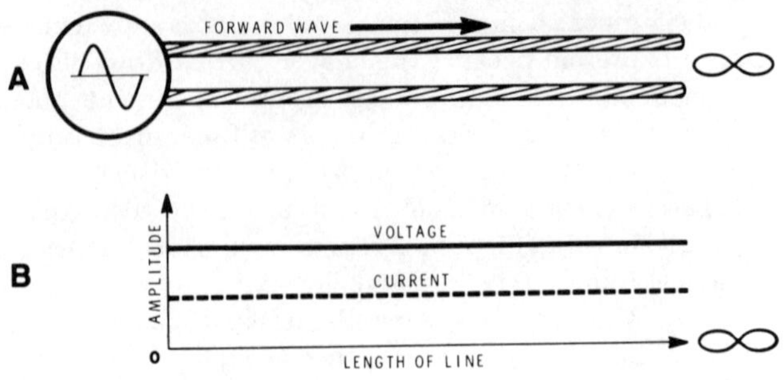

Figure 6-3
An infinite transmission line.

Since voltage (E) and current (I) are constant at all points, impedance (Z) is also constant, proven by the Ohm's law equation($Z = \frac{E}{I}$). Therefore, the infinite length of transmission line has a constant impedance. This is known as its **characteristic impedance**. Every transmission line has a specific characteristic impedance which is due to its physical construction. The formula for the characteristic impedance of a balanced transmission line is:

$$Z_0 = 276 \times \log \frac{d}{r}$$

where Z_0 = characteristic impedance in ohms

d = center-to-center distance of separation of conductors

r = radius of conductors (using the same unit of measurement as with d)

Figure 6-4 shows a balanced transmission line with measurements d and r indicated.

The following example shows how to find the characteristic impedance of a typical balanced transmission line. The distance between conductors is 1.22 cm and the conductor radius is 0.1 cm. Using the formula for characteristic impedance:

$$Z_0 = 276 \times \log \frac{d}{r}$$

$$Z_0 = 276 \times \log \frac{1.22}{0.1}$$

$$Z_0 = 276 \times \log 12.2$$

$$Z_0 = 276 \times 1.086$$

$$Z_0 = 300\ \Omega$$

Thus, the characteristic impedance of this particular line is 300 Ω.

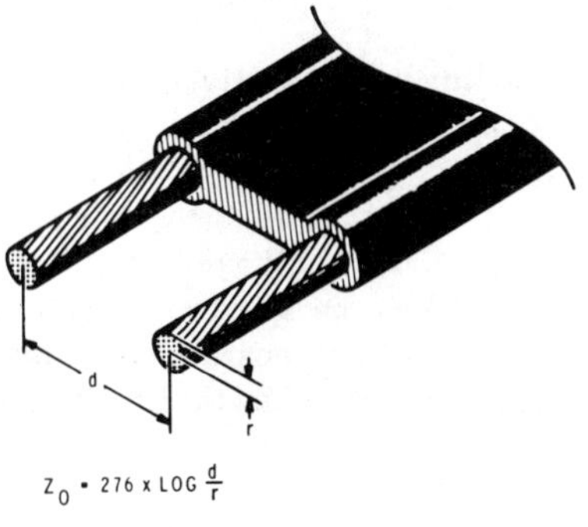

Figure 6-4
Balanced transmission line dimensions.

If you connect an infinitely long length of this 300 Ω line to an RF generator, the input impedance that the generator will see is 300 Ω. In fact, the impedance at all points on the line is 300 Ω. Therefore, if you cut the line to some practical length and connect a 300 Ω resistor across the far end of the line, the generator will still see a 300 Ω input impedance. You are, in essence, replacing the lost portion of the line with a 300 Ω resistor. Since the resistor has the same impedance as the lost portion of the line, the previous conditions are not upset. Therefore, the transmission line behaves just as though it were infinitely long. The forward traveling wave of RF energy travels the length of the line and is completely dissipated in the 300 Ω resistor. This condition exists whenever a transmission line is correctly terminated; that is, whenever the load impedance matches the characteristic impedance of the transmission line. In the example given, the 300 Ω load resistor matched the 300 Ω transmission line.

Standing Waves

If a transmission line has infinite length or is terminated in its characteristic impedance, all the power applied to the line at one end is absorbed or dissipated by the load at the other end. However, if a transmission line is terminated in an impedance other than its characteristic impedance, some (but not all) of the applied power is absorbed by the load. The remaining power is **reflected**.

When a transmission line is incorrectly terminated, the power that is not absorbed by the load is sent back toward the generator. The amount of this reflected power is directly proportional to the amount of impedance mismatch at the load. For example, if the line is terminated by a short circuit or an open circuit, none of the power is dissipated in the termination and all of it is reflected back to the generator. As the termination impedance approaches the line's characteristic impedance, less and less power is reflected, until finally, when the impedances match, no power is reflected.

When power is applied to a transmission line, a voltage and current appear whose values depend on the characteristic impedance (Z_O) and the applied power. The voltage and current waves travel to the load (Z_L) and, if $Z_L = Z_O$, the load absorbes all the power and none is reflected. The only waves present at this time are the forward traveling waves of current and voltage from the generator to the load.

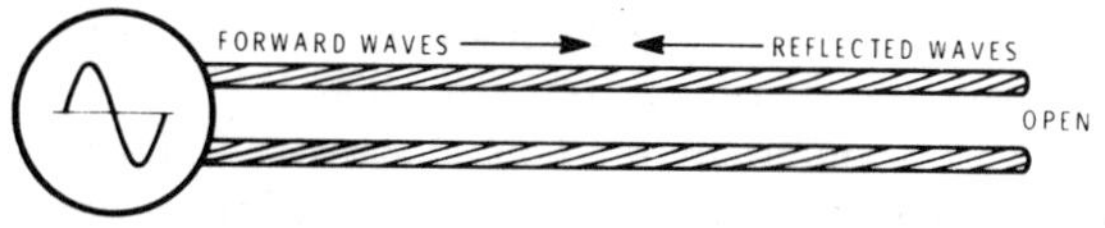

Figure 6-5
Forward and reflected waves.

However, if Z_L is not equal to Z_0, some power is absorbed and the rest is reflected. Therefore, one set of current and voltage waves is traveling toward the load and another set is traveling back to the generator. This is shown in Figure 6-5. These two sets of traveling waves, going in opposite directions, will alternately add to and subtract from one another due to their relative phase differences. This in turn sets up a pattern known as **standing waves** along the line. The standing wave pattern for an open-circuited line is shown in Figure 6-6. You can see that **stationary** voltage and current minimums and maximums have appeared. Compare this voltage and current pattern to that of a correctly terminated transmission line shown in Figure 6-3. In the infinite or correctly terminated line, there are no reflected waves to interfere with the forward waves. How do these standing waves develop?

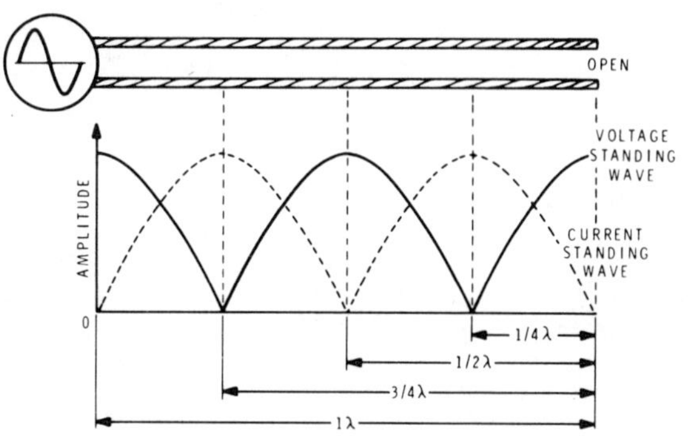

Figure 6-6
Standing wave pattern for an open circuited line.

For the moment, consider only the forward traveling voltage and current waves. At the load, the current is zero and the voltage is maximum because the load is an open circuit. This is shown in Figure 6-6. The reflection that occurs at the open circuit affects both voltage and current. The voltage now starts traveling back to the generator, unchanged in phase, but the current is reflected with a 180° phase inversion. Therefore, at a point exactly a quarter-wavelength (1/4 λ) from the load, the voltage is permanently zero as shown in Figure 6-6, because the forward and reflected voltage waves are exactly 180° out of phase. This occurs because the reflected wave has had to travel a distance of 1/4 λ + 1/4 λ or 1/2 λ farther than the forward wave. Since 1/2 λ is equal to a phase change of 180°, the two waves cancel and voltage is zero at this point. The current wave has also traveled an extra 1/2 λ; however, since it underwent a 180° phase inversion on reflection, its total phase change is 360°. Therefore, because the forward and reflected current waves are in phase, reinforcement takes place and a current maximum occurs at exactly the same point as the voltage minimum.

Figure 6-6 shows that 1/2 λ from the load, there is a current minimum (zero in this case) and a voltage maximum. This occurs because the forward and reverse voltage waves are now in phase, voltage has traveled a distance of one wavelength (360°) to return to this point. At this same place, the current waves cancel, because of the additional 180° phase inversion that occurred on reflection. Figure 6-6 shows that these maximum and minimum points repeat every half-wavelength and reverse every quarter-wavelength.

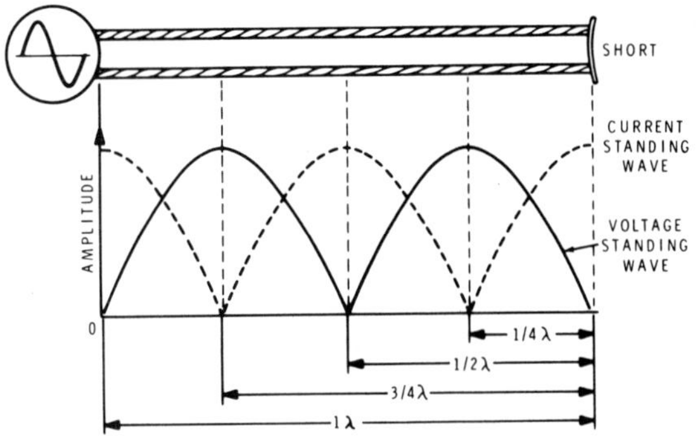

Figure 6-7
Standing wave pattern for a short circuited line.

Note that this condition is permanent and is determined by the load. Figure 6-7 shows the standing-wave pattern for a short-circuited transmission line. Here, the current at the load is maximum and the voltage is zero because the load is a short circuit. At a quarter-wavelength from the load, current is zero and voltage is maximum. At a half-wavelength from the load, current is maximum and voltage is zero. Thus, the standing-wave pattern for a short-circuited line is just the opposite of that for an open-circuited line. This shows that the load does indeed determine the standing-wave pattern.

Standing Wave Ratio

The ratio of maximum current to minimum current or maximum voltage to minimum voltage along a transmission line is called the standing-wave ratio (SWR). This ratio is a measure of the mismatch between the load and the transmission line. For example, the SWR is equal to 1 when the load is perfectly matched, and equal to ∞ when the load is a short or open circuit. If the transmission line is terminated in a purely resistive load, the standing-wave ratio is found using the formula:

$$SWR = \frac{Z_O}{R_L} \text{ or } SWR = \frac{R_L}{Z_O}$$

(Whichever is greater than 1)

Where R_L = Load resistance

Z_O = Characteristic impedance

The standing-wave ratio is always equal to or greater than one. Regardless of whether R_L is half as large or twice as large as Z_O, the ratio of a voltage maximum to a voltage minimum is 2:1. Therefore, the degree of mismatch is the same in both instances.

The higher the SWR is, the greater is the mismatch between line and load. Also in practical transmission lines, power loss increases with SWR. Therefore, a low value of SWR is always sought, except when the line is used as a tuned or resonant circuit.

Transmission Line Resonant Circuits

A transmission line that is terminated in its characteristic impedance has an SWR of 1 and is called a **non-resonant** or flat transmission line. It is non-resonant because it presents a constant load impedance to the generator regardless of the operating frequency. On the other hand, a transmission line with an SWR greater than 1 presents a variable load impedance that is dependent on the operating frequency. Therefore, a line with an SWR greater than 1 is called a **resonant** transmission line.

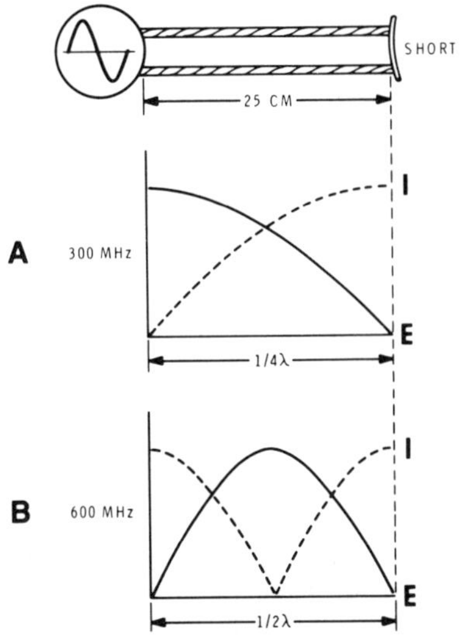

Figure 6-8
A resonant transmission line.

Figure 6-8 shows a 25 cm length of transmission line that is terminated with a short circuit. Therefore, the SWR is infinite. Figure 6-8A shows the standing-wave pattern that develops on the line when a 300 MHz signal is applied. The 25 cm length of line is 1/4 λ at 300 MHz because

$$\lambda\,(\text{meters}) = \frac{3 \times 10^8 \text{ meters/second (velocity of wave)}}{\text{frequency (Hz)}}$$

$$\lambda = \frac{3 \times 10^8}{300 \text{ MHz}} = \frac{3 \times 10^8}{3 \times 10^8 \text{ Hz}} = 1 \text{ meter}$$

Therefore, 1/4 λ is 0.25 m or 25 cm at 300 MHz. Notice that the input of the line in Figure 6-8A has maximum voltage and minimum or zero current. This corresponds to an open circuit and it is in fact the input impedance of the 1/4 λ shorted transmission line. The quarter-wave line actually inverted the impedance of the load. If the load had been an open circuit, the input impedance would have been a short circuit. Since these conditions are only true when the line is exactly 1/4 λ long, if the input frequency changes, so does the input impedance. Because the transmission line is frequency dependent, it is said to be resonant. In fact, the 1/4 λ shorted line acts exactly like a parallel resonant circuit. That is, at its resonant frequency or the frequency at which it is exactly 1/4 λ long, its input impedance is infinite or open. At all other frequencies, its input impedance becomes either capacitive or inductive depending on whether it is above or below its resonant frequency.

Transmission lines are used as both parallel and series resonant circuits in VHF, UHF, and microwave equipment. They not only offer an extremely high Q but often are the only practical method of obtaining a resonant circuit at these high frequencies. This is because, as frequency increases, the inductors and capacitors required become smaller and smaller. When this happens, circuit Q drops and bandwidth increases. The only way to improve this is to use transmission line resonant circuits.

To be sure you understand this important concept, consider Figure 6-8B. Here is the same 25 cm length of transmission line. However, since the input frequency has been doubled to 600 MHz, the line appears as a half-wavelength. This accounts for the different standing-wave pattern. Note that the input now has maximum current and minimum or zero voltage. Therefore, the input is a short circuit for a 1/2 λ shorted transmission line. Since this only occurs at the frequency at which the line is exactly 1/2 λ, it acts as a resonant circuit. In fact, the 1/2 λ shorted line acts like a series resonant circuit. Notice that the 1/2 λ line duplicates the terminating impedance at its input. Therefore, the input of a 1/2 λ open transmission line will also be an open.

To summarize, a 1/4 λ line inverts its terminating impedance. Therefore, the input of a shorted 1/4 λ line is an open and the input of an open 1/4 λ line is a short. Thus, a shorted 1/4 λ can be used as a parallel resonant circuit and an open 1/4 λ line can be used as a series resonant circuit. Conversely, a 1/2 λ line duplicates its terminating impedance. Therefore, the input of a shorted 1/2 λ line is a short and the input of an open 1/2 λ line is an open. Thus, a shorted 1/2 λ line can be used as a series resonant circuit and an open 1/2 λ line can be used as a parallel resonant circuit.

A correctly terminated transmission line has a constant input impedance regardless of frequency or line length. For example, if a 300 Ω transmission line is terminated by a 300 Ω resistor, its input impedance will be 300 Ω no matter what the frequency or the length of the line. It is for this reason that a matched condition, indicated by an SWR of 1, is always sought when installing a transmission line. The only time a high SWR is desired is when the line is being used as a resonant circuit as shown previously.

Figure 6-9
Reflectometer or SWR meter.

SWR Measurement

Standing-wave ratio is measured by a **reflectometer** or an SWR meter. A typical SWR meter is shown in Figure 6-9. This unit measures the ratio between forward and reflected power and displays it on a calibrated meter.

Figure 6-10 shows that the SWR meter is connected in series in the transmission line, between the transmitter and the load or antenna. SWR is measured by first tuning the transmitter for maximum output. The SWR sensitivity control is then set for full-scale meter deflection. The meter is then switched to measure SWR and the standing-wave ratio is read directly.

The meter is actually indicating forward power when the SWR sensitivity control is set for full scale. Then the meter is switched to measure reflected power. The amount of reflected power is indicated; however, the meter is calibrated to read the standing-wave ratio directly. This is a quick and easy method of checking SWR, with a relatively low-cost instrument.

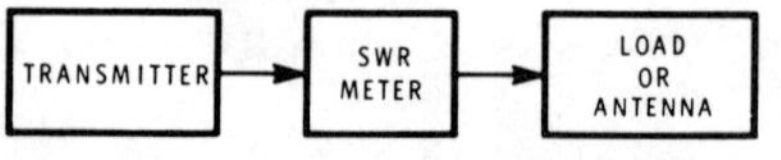

Figure 6-10
An SWR meter is connected in series in the line between transmitter and load.

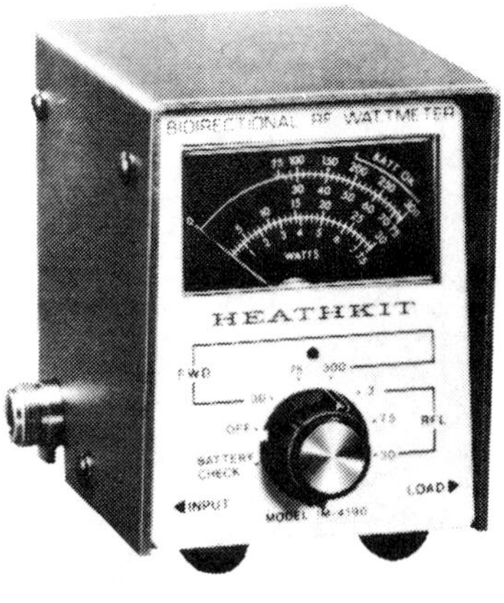

Figure 6-11
A directional wattmeter.

A meter that measures the actual amount of reflected and forward power is shown in Figure 6-11. It is a directional wattmeter and, with the flick of a switch, you can measure forward and reflected power. A perfectly matched condition, SWR = 1, is indicated by zero reflected power. For other values of reflected power, the standing wave ratio is determined from the chart shown in Figure 6-12. You can use this chart to find the SWR when forward and reflected power are known. For example, if the forward power is 5 watts and reflected power is 0.2 watts, find these lines on the chart; the point where they intersect indicates the standing-wave ratio. In this case, they intersect on the diagonal line that represents an SWR of 1.5.

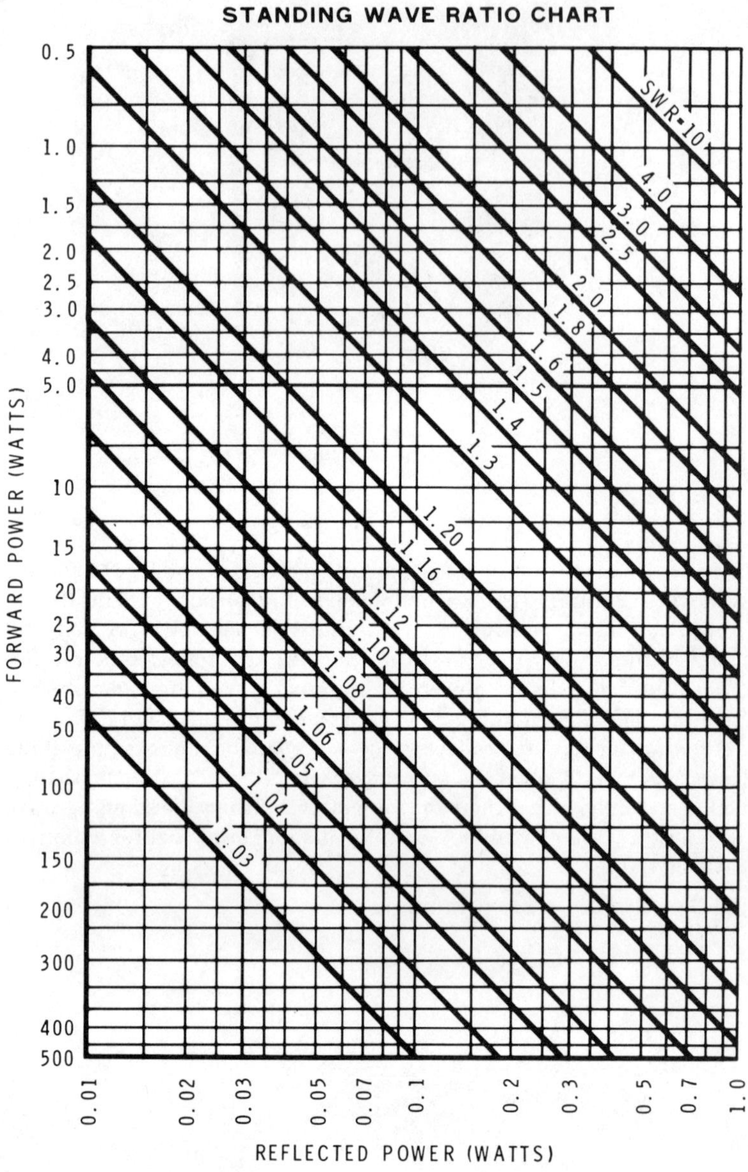

Figure 6-12
Standing wave ratio chart.

Coaxial Transmission Lines

Figure 6-13A shows a coaxial type of transmission line, which is one of the most commonly used types of lines. It is an **unbalanced** line since it consists of a shield that is at ground potential, and a center conductor. The center conductor "carries" the RF energy and the shield prevents this energy from being radiated into space. Thus, losses due to radiation are kept to a minimum. With a balanced line, radiation is minimal due to the balanced currents. The radiated fields created by each line are 180° out of phase and, therefore, cancel.

A coaxial line has a specific characteristic impedance just as a balanced line does. The formula for the characteristic impedance of a coaxial line is:

$$Z_0 = 138 \times \log \frac{D}{d}$$

Where Z_0 = characteristic impedance in ohms.

D = inside diameter of shield.

d = outside diameter of the center conductor.

These dimensions are shown in Figure 6-13B.

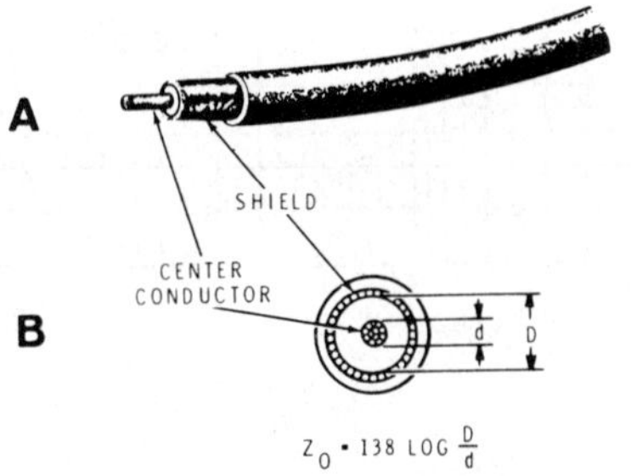

Figure 6-13
Coaxial transmission line.

Comparison of Coaxial and Balanced Lines

When transmission line characteristics were discussed earlier, the lines were assumed to be perfect. However, practical transmission lines exhibit power losses. These losses are due to the resistance of the conductors, leakage in the dielectric material that separates the conductors, and in some cases, radiation. Figure 6-14 shows the normal attenuation of three typical transmission lines versus the operating frequency. RG-58/U cable is a small diameter 52 Ω coaxial cable, RG-8/U cable is a larger diameter 52 Ω coaxial cable, and 214-022 cable is a 300 Ω balanced transmission line. You can see that the balanced transmission line offers the lowest attenuation; however, this advantage is offset to some degree by its installation requirements.

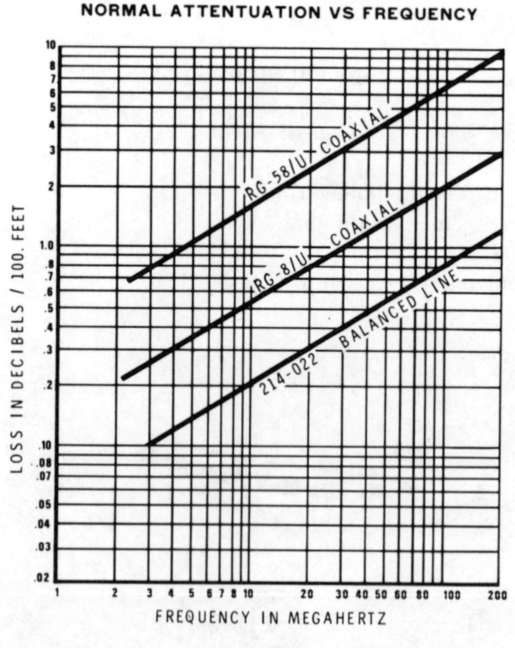

Figure 6-14
Attenuation in transmission lines.

When you install a balanced transmission line, you must prevent it from collecting moisture. Any moisture that accumulates on the line will change the dielectric between the two lines, which changes its characteristic impedance. The result is reflections and additional line losses. You must also keep the line clear of other conductors such as downspouts, rain gutters, and metal towers. The line must be spaced at least two to three times the conductor separation from any other conductors. For example, if the distance between the two wires of the transmission line is 1/2", the line must be spaced 1" to 1-1/2" away from other conductors. This requires standoff insulators.

Coaxial line, on the other hand, has an outside shield that is at ground potential, so you can mount it anywhere. You can tape it directly to any metal object without adverse effects. Although moisture can effect coaxial cable if it enters the cable, it has little effect when proper care is taken to prevent moisture buildup on the inside of the cable. Thus, while balanced line offers the lowest attenuation, coaxial cable is widely used because it's easy to install.

ANTENNAS

After a transmitter generates an RF signal, there must be some method of radiating this signal into space. There must also be some method for a receiver to intercept or "pick up" the signal. The **antenna** fulfills both requirements.

An antenna, generally made of metal, (often a wire or collection of wires) converts high frequency current into electromagnetic waves for transmission and does just the opposite for reception. Transmitting and receiving antennas have different functions, but they behave identically. That is, their behavior is reciprocal.

Antenna Principles

Figure 6-15 shows a 1/4 λ open-circuited transmission line. You can see that the forward and reflected waves combine to form a standing-wave pattern on the line. This was discussed previously, but it wasn't mentioned that not all the forward energy is reflected by the open circuit. Actually, a small portion of the RF energy escapes from the system and is radiated. This happens because the RF energy traveling toward the open circuit is required to undergo a violent change, a phase reversal, when it reaches the open. A small portion of the forward wave cannot do this, and it actually "shoots" past the open circuit into free space. The amount of waves escaping the transmission line is very small, for two reasons. First, if the surrounding space is considered as a load for the transmission line, a mismatch exists, and very little power is dissipated in this "load." Second, since the two wires are close together and 180° out of phase, the radiation from one will cancel that from the other.

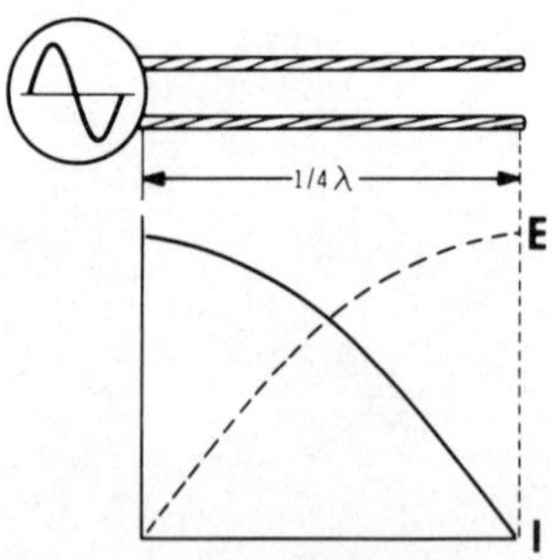

Figure 6-15
Open circuited quarter wave transmission line.

To develop this "open" into a better radiator, we must "enlarge" the open circuit, or spread the two wires. When the wires are spread out, the radiation is less likely to cancel. Also, the radiating transmission line is better coupled to the surrounding space because more power is "dissipated" or radiated. Moreover, because the wires are spread out, waves traveling along the line find it increasingly difficult to reverse phase. Thus, everything points to an increase in radiation.

Radiation can be increased even further by spreading the wires until they are in line, as shown in Figure 6-16A. The electrostatic or electric field and the electromagnetic field of the line are now fully coupled to the surrounding space. Therefore, the maximum amount of radiation results. This type of antenna is called a **dipole**. The original transmission line was 1/4 λ long, but when it is spread out, the overall length is 1/2 λ. Therefore, the antenna shown in Figure 6-16A is a **half-wave dipole**.

The current and voltage standing-wave patterns on the half-wave dipole are shown in Figure 6-16. Notice that both ends appear as opens and, therefore, have a voltage maximum and current minimum. The center of the antenna, which is the feed point, has a current maximum and voltage minimum. As a result, the input impedance of the half-wave dipole is low. While the voltage and current distribution indicates that the input impedance should be zero ohms, in actuality, it is 73 ohms. This is due to the energy lost to radiation. This radiated energy is not reflected back to the input and, therefore, complete cancellation does not occur. For this reason, input impedance rises to 73 ohms as the 1/4 λ open line is spread out into a dipole.

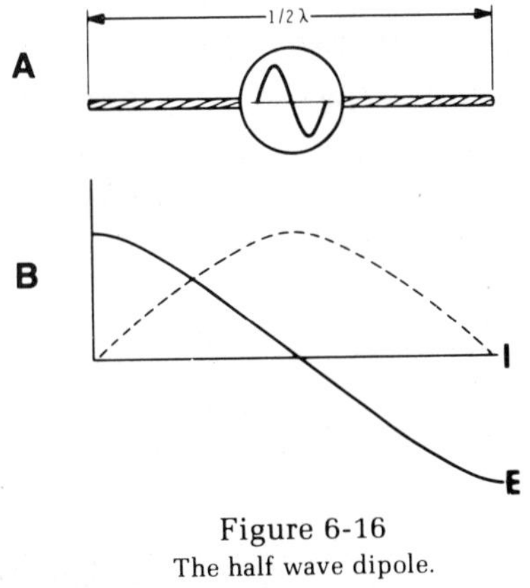

Figure 6-16
The half wave dipole.

Antenna Radiation

Since a dipole has high voltage potentials at both ends, an electrostatic or **electric field** exists between these points. This field is just like the one that exists between the plates of a capacitor. However, in the case of an antenna, the electric field is not confined to the area between the plates; instead, it is radiated into space. The electric field around a half-wave dipole is shown in Figure 6-17.

Note also from Figure 6-17 that a **magnetic field** exists around the antenna. This is due to the current flow in the antenna. Since current is maximum at the center of the antenna, so is the magnetic field. Both the magnetic and electric fields are radiated as an **electromagnetic wave**. This wave propagates or travels outward from the antenna and will continue traveling even after the current and voltage are removed from the antenna.

The farther the electromagnetic wave travels from the antenna, the weaker it becomes. The amplitude of this wave or radiation field is evaluated in terms of the voltage it will induce across a wire. This is known as its **field strength**.

The field strength at any location depends on the distance from the transmitter and on the radiated power. The strength varies inversely with distance. For example, if the distance from the antenna is doubled, the field strength will be reduced by one-half. Conversely, if the distance is cut in half, the field strength will be doubled. Likewise, if the radiated power is increased, the field strength at any location will also increase. However, since field strength is indicated by the voltage induced in a wire, any power increase must be converted to a corresponding voltage increase. Therefore, since power is proportional to voltage squared ($P = \frac{E^2}{R}$), it follows that the increase in field strength is proportional to the square root of the power change. Thus, doubling the radiated power will increase the field strength by $\sqrt{2}$ or 1.414. Conversely, to increase the field strength by 2, it is necessary to increase the radiated power by 4.

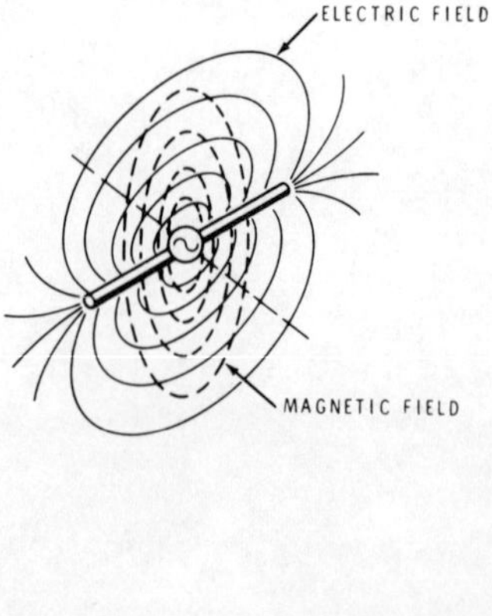

Figure 6-17
Magnetic and electric field around a half wave dipole.

Antenna Polarization

As discussed earlier, the electromagnetic wave radiated from any antenna has two component fields: the electric field and the magnetic field. These fields are perpendicular to each other and both are perpendicular to the direction of propagation or travel. This is shown with vectors in Figure 6-18. In this case, the electric field vector is horizontal. Therefore, the wave is said to be **horizontally polarized**. If the pattern were rotated 90°, the electric vector would be vertical and the wave would be vertically polarized. Thus, the electric field vector determines the polarization of the wave. As you will see later, a horizontal antenna produces horizontal polarization, while a vertical antenna produces vertical polarization. The importance of this is that a horizontally polarized wave will induce maximum voltage in a horizontal antenna. Theoretically, a horizontally polarized wave will induce zero volts in a vertical antenna. However, in practice, at HF and VHF, this seldom occurs because of wave changes that take place during propagation.

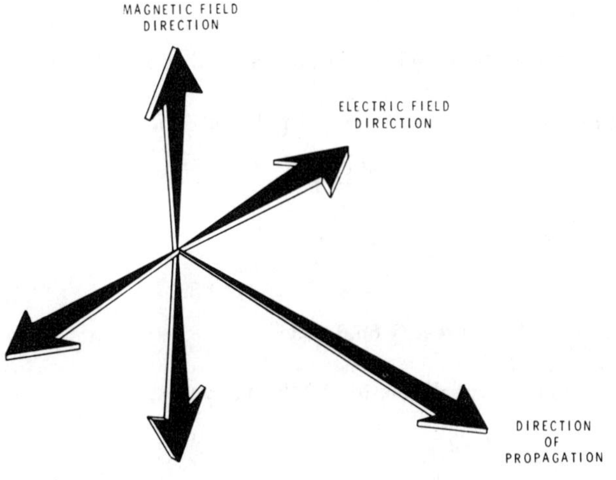

Figure 6-18
Polarization of an electromagnetic wave.

The Dipole Antenna

The most commonly used type of antenna is the half-wave dipole. As its name implies, it is 1/2 λ long at the operating frequency. The formula for wavelength is:

$$\lambda = \frac{3 \times 10^8}{f}$$

Where λ = wavelength in meters

f = frequency in hertz

This formula gives the wavelength in free space. However, with an antenna, the wave is developed on a wire and **end effects** must be considered. End effects make an antenna appear electrically as though it were about 5% longer than its actual physical length. This is due to a capacitance that exists between the ends of the antenna. The formula for finding the length of a half-wave dipole that is supported by insulators at each end is:

$$1/2\ \lambda \text{ in feet} = \frac{468}{f_{MHz}}$$

This formula gives the length in feet and takes end effects into account.

Find the length of a half-wave dipole for 7 MHz.

$$1/2\ \lambda \text{ in feet} = \frac{468}{f_{MHz}}$$

$$1/2\ \lambda \text{ in feet} = \frac{468}{7}$$

$$1/2\ \lambda \text{ in feet} = 66.9 \text{ feet}$$

Thus, a half-wave dipole for 7 MHz is 66.9 feet long.

A typical dipole installation is shown in Figure 6-19. Notice that the antenna is fed at the center with a 75 Ω balanced transmission line. This gives a good match to the antenna's 73 Ω input impedance.

A dipole can also be "fed" with a coaxial transmission line. The center conductor is connected to one side of the dipole and the shield to the other side.

However, since the coaxial line is unbalanced and the dipole balanced, having been developed from a 1/4 λ balanced line, some inefficiency will result. The current and voltage distribution on the antenna will be upset and RF current will actually flow on the coaxial shield. This shield current may result in radiation from the cable, which is undesirable.

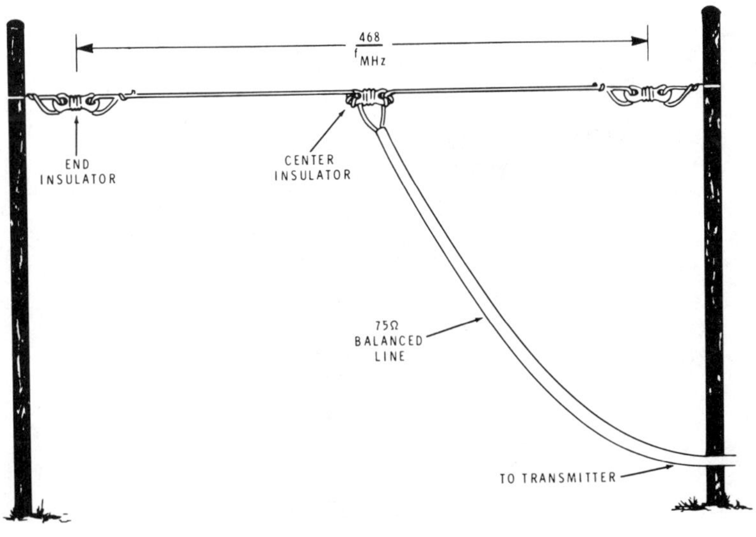

Figure 6-19
Half wave dipole installation.

To correctly couple a coaxial line to a dipole antenna, a **balun** is used. A balun is a **bal**anced to **un**balanced RF transformer. A typical balun is shown in Figure 6-20. This particular balun also acts as a center insulator for the dipole. As shown, the wires connect to the dipole and the connector is for the coaxial cable. Baluns are constructed for specific impedance ratios, just like conventional transformers. When 50 Ω or 75 Ω coaxial cable is used to feed a dipole, an impedance ratio of 1:1 is used.

By far, the most important property of an antenna is its pattern of radiation. In the case of a transmitting antenna, this is a graphical plot of the field strength radiated by the antenna in different angular directions. This same radiation pattern also indicates the receiving properties of the antenna. This is because the receiving and transmitting properties of an antenna are reciprocal.

Figure 6-20
A 1:1 balun. Photo courtesy Palomar Engineers.

Figure 6-21A shows a 3-dimensional view of the radiation pattern for a horizontal half-wave dipole. Notice that maximum radiation occurs broadside to the antenna and minimum radiation occurs off the ends of the antenna. A polar plot of the radiation in the horizontal plane is shown in Figure 6-21B. The radiation pattern for a vertical half-wave dipole is shown in Figure 6-22A. Note that, in this case, maximum radiation occurs in all directions in the horizontal plane. Minimum radiation occurs directly above and below the antenna. The polar plot for the horizontal plane is shown in Figure 6-22B.

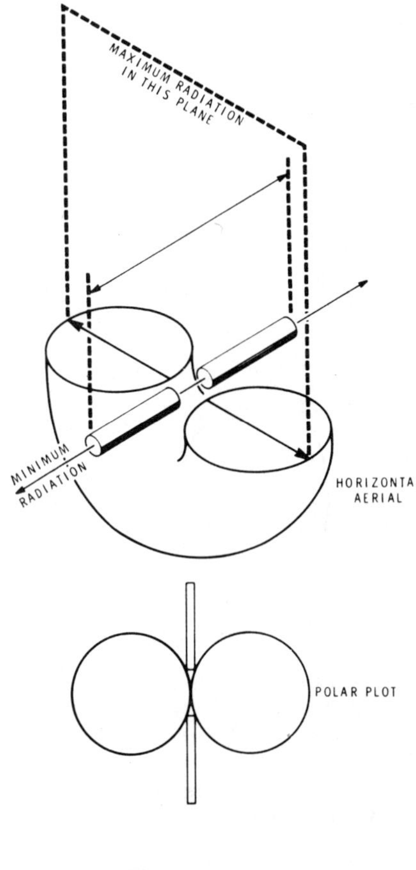

Figure 6-21
Horizontal half-wave dipole radiation pattern.

A

B

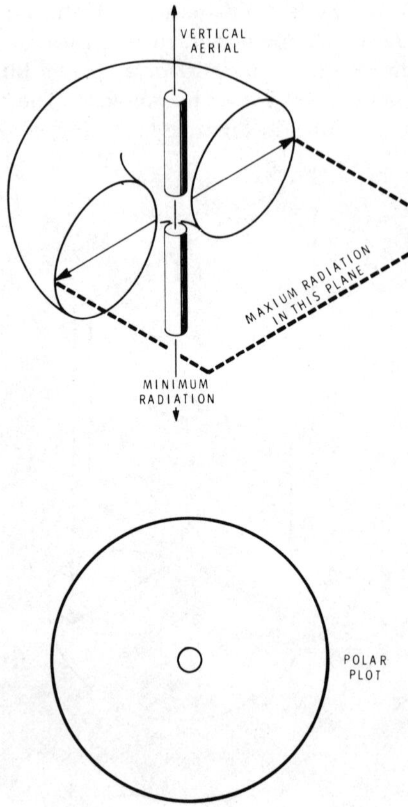

Figure 6-22
Vertical half wave dipole radiation pattern.

A dipole antenna can also operate on harmonics of its fundamental frequency. That is, a half-wave dipole can operate on its second harmonic, where it functions as a full-wave dipole. The current distribution for a full-wave dipole is shown in Figure 6-23A. Note that minimum current occurs at the center of the antenna. This indicates that the center of the antenna is a high impedance point. If a low input impedance is desired, the antenna feedpoint must be moved to either of the current maximums which are located 1/4 λ from either end. The radiation pattern for a full-wave dipole is shown in Figure 6-23B. Note that the pattern has split up into four major lobes. This is due to the current distribution on the dipole.

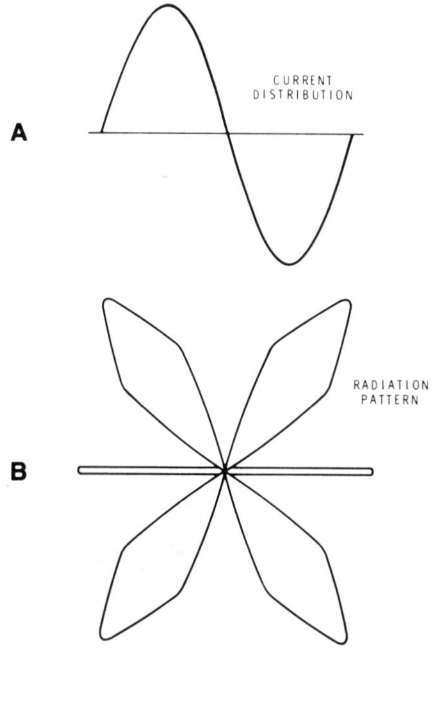

Figure 6-23
The full wave dipole.

A half-wave dipole can also operate on its third harmonic as shown in Figure 6-24. The current distribution for the resulting 1-1/2 λ antenna is shown in Figure 6-24A. Note that, here, a current maximum occurs at the center of the antenna. Therefore, the low impedance feedpoint remains at the center of the antenna. The radiation pattern for the 1-1/2 λ antenna is shown in Figure 6-24B. Note that there are still four major lobes, but now, there is an additional minor lobe broadside to the antenna.

The harmonic operation of dipoles can be put to good use by designing an antenna for use on two harmonically related frequencies. A good example is an antenna designed for the 7 MHz and 21 MHz amateur radio bands. It is cut to a half-wavelength at 7 MHz. Thus, when operated at 21 MHz, it operates as a 1-1/2 wavelength antenna. Also, since both the 1/2 λ and 1-1/2 λ antennas have a low impedance center feedpoint, due to a current maximum at the center, they offer a good impedance match to a 75 Ω or 50 Ω transmission line.

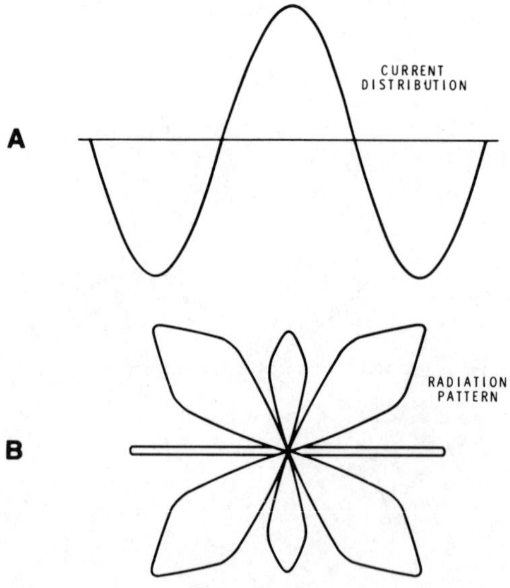

Figure 6-24
1-1/2 wave dipole.

The Vertical Antenna

In cases where vertical polarization is required, the antenna must be vertical. However, at low frequencies, the height of a vertical half-wave dipole can become prohibitive. For example, at 3.5 MHz, a vertical half-wave dipole would be 134 feet high and at 1.8 MHz, it would be 263 feet high.

This would seem to indicate that a low frequency vertical antenna would be impractical. However, if a quarter-wavelength vertical antenna is constructed above a perfect ground, it will have the same characteristics as the half-wave vertical dipole. This is possible because a perfect ground will produce a "mirror image" of the quarter wave, thus, producing the effect of a half-wave antenna. The image results from the reflected radio waves as shown in Figure 6-25A. The voltage and current distribution for a quarter-wave vertical is shown in Figure 6-25B. In order for this antenna to operate correctly, the ground must be a perfect conductor. If it is not, considerable power will be lost in the resistance of the grounding system. When an antenna is constructed over soil of poor conductivity, such as rocky or sandy soil, an artificial ground system must be used. This consists of several 1/4 λ copper wires extending outward from the base of the antenna in as many directions as possible. These wires are referred to as **radials** and usually four radials are the bare minimum. When only four radials are used, they should be perpendicular to each other.

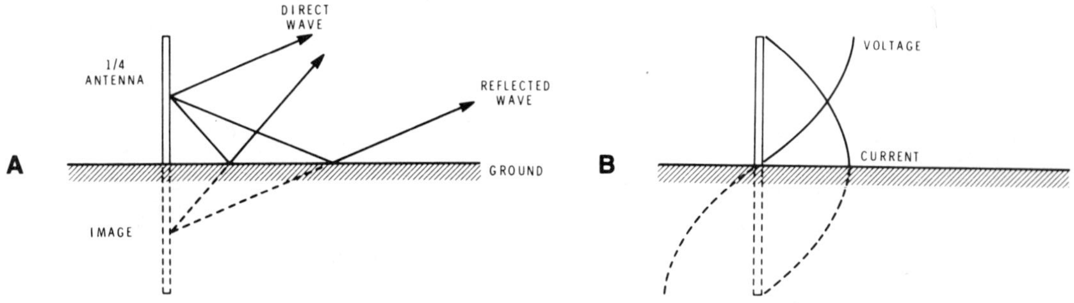

Figure 6-25
The quarter wave vertical.

The radiation patterns for a quarter-wave vertical antenna are shown in Figure 6-26. In the horizontal plane, the antenna radiates equally well in all directions. It is omnidirectional and gives good "coverage" in all directions. In the vertical plane, the radiation is directed low toward the horizon. This gives it excellent, long distance propagation characteristics. Wave propagation is discussed in detail in the next section.

The input impedance of the quarter-wave antenna is around 36 Ω when it is used with a perfect ground. This is a good match for a 50 Ω coaxial transmission line. Also, since the vertical antenna is unbalanced because one side is ground, unbalanced coaxial cable is a perfect feeder. The center conductor is connected to the quarter-wave antenna, which is insulated from ground, and the shield is connected to ground or the radial system.

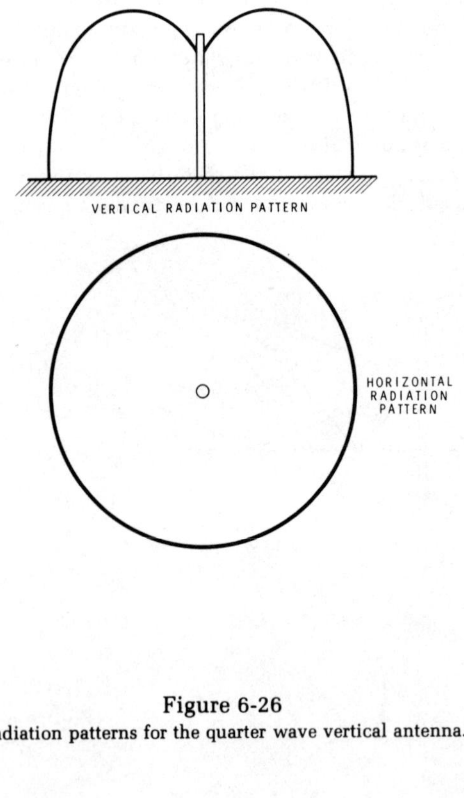

Figure 6-26
Radiation patterns for the quarter wave vertical antenna.

Antenna Arrays

In many situations, the direction of the radiated wave must be restricted within some specific limits. This is done to eliminate or reduce interference to or from other stations. It also results in more efficient operation, since any power radiated is directed towards the specific receiving station. Therefore, very little (if any) power is wasted.

When the radiation pattern of an antenna system is essentially in one direction, it is unidirectional. Contrast this with the half-wave horizontal dipole which is bi-directional and the quarter wave vertical which is omnidirectional. The measure of a unidirectional antenna's directional properties or "directivity" is its **beam width**. This is measured from the radiation pattern as the angle between the two points on either side of maximum radiation where the field strength drops 3 dB. A typical unidirectional radiation pattern is shown in Figure 6-27. In this case, the beam width is 50°.

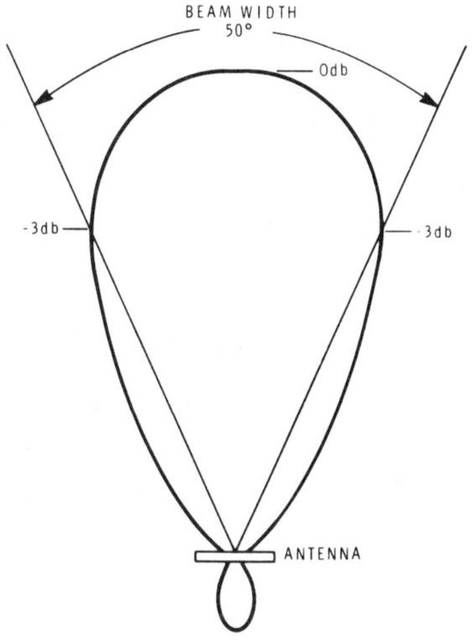

Figure 6-27
Measuring bandwidth of a unidirectional antenna.

An antenna system with good directivity or narrow beam width has the advantage of providing directional **gain**. Since the radiated power is concentrated within a relatively small beam, the field strength within this beam is much greater than would be obtained from an omnidirectional antenna. Therefore, the **effective radiated power** (ERP) of the transmitter is increased by the directional gain of the antenna. **Directional gain** is the ratio of the power required to produce a given field strength at a given location using a reference antenna, compared to the power required to produce the same field strength with a directional antenna. Reference antennas used are usually half-wave dipoles or quarter-wave vertical antennas. Very high gains are possible, particularly at VHF and UHF where electrically large antenna systems can be built in a relatively small space.

Effective radiated power or ERP can be calculated if transmitter output power, transmission line loss, and antenna gain are known. For example, if the transmitter output power is 100 W, transmission line loss is 10 W, and antenna power gain is 10; ERP is:

100 W	Transmitter output
− 10 W	Line loss
90 W	Input to antenna
× 10	Antenna power gain
900 W	Effective radiated power

This shows how a directional antenna greatly increases the effective radiated power.

Directivity or directional gain is obtained by using combinations of two or more antenna elements to form an **antenna array**. Depending on the method used to excite the additional elements, antenna systems are classified as **parasitic** or **driven arrays.**

PARASITIC ARRAYS

An antenna element that is not connected to the transmission line will develop a voltage through induction. It is called a **parasitic element**. Figure 6-28 shows a parasitic element located 1/4 λ from a dipole, which is called the **driven element**. Both elements are 1/2 λ long and, therefore, are resonant at the operating frequency. The radiation pattern of the dipole, by itself, is bidirectional and is maximum broadside to the antenna. If you consider the energy traveling toward the parasitic element, it travels 1/4 λ. Therefore, before it reaches the parasitic element, it has gone through a 90° phase change. As the wave cuts the parasitic element, a voltage is induced that is 180° reversed with respect to the wave that induced it. As a result, current flows through the element and the element radiates. In the direction beyond the parasitic element, the field radiated by it is opposite to the field from the driven element. Therefore, the two fields cancel, and radiation in this direction is negligible. However, the parasitic element also radiates toward the driven element. By the time this field reaches the dipole, it has gone through a further 90° phase change. Therefore, the total phase change is 360°, and the wave arrives in phase with the energy from the driven element. Thus, radiation toward the driven element is reinforced and maximum radiation results. The direction of maximum radiation is called the forward direction, and the parasitic element is called the **reflector**.

The radiation pattern obtained with parasitic arrays depends on the magnitude and phase of the current in the parasitic elements. These factors depend on the length of the parasitic elements and on the spacing between the driven and parasitic elements. Optimum reflector operation is obtained with a spacing of 0.18 to 0.2 λ and with an element length approximately 5% longer than 1/2 λ. Also, if the parasitic element is approximately 5% shorter than 1/2 λ and the spacing is reduced to about 0.1 λ radiation in the direction of the parasitic element will be reinforced. In this case, the element is called a **director**.

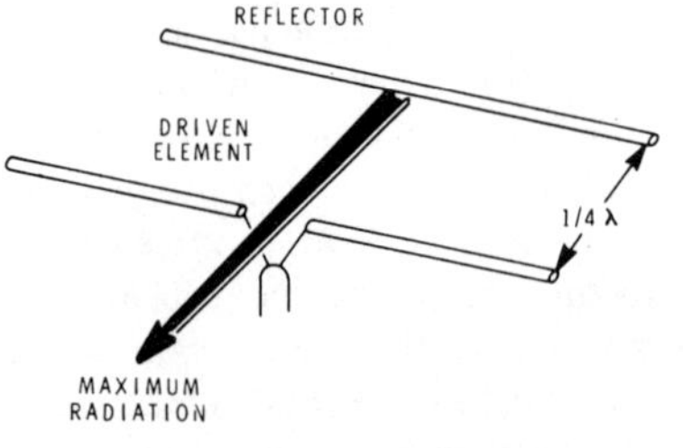

Figure 6-28
A half wave dipole driven element with a parasitic reflector.

A 3-element parasitic array is shown in Figure 6-29A. This is known as a Yagi antenna, and it is named after one of its inventors. Figure 6-29B shows the radiation patterns of the driven element alone, with the reflector added, and with the director added. You can see that the directivity improves with each additional element. Still sharper directivity and therefore, increased gain can be achieved by adding more directors. An example is shown in Figure 6-30. This is a 6-element Yagi designed for 440 MHz. Notice that it is oriented for vertical polarization.

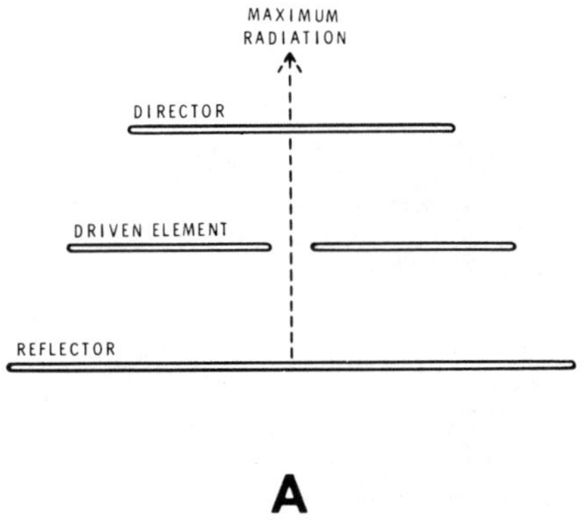

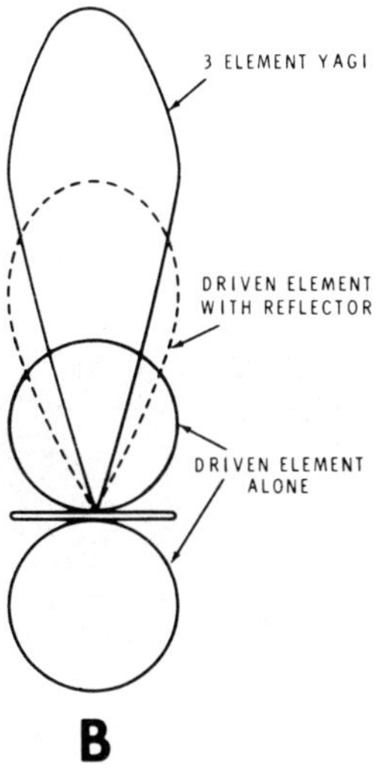

Figure 6-29
A three-element Yagi and its radiation pattern.

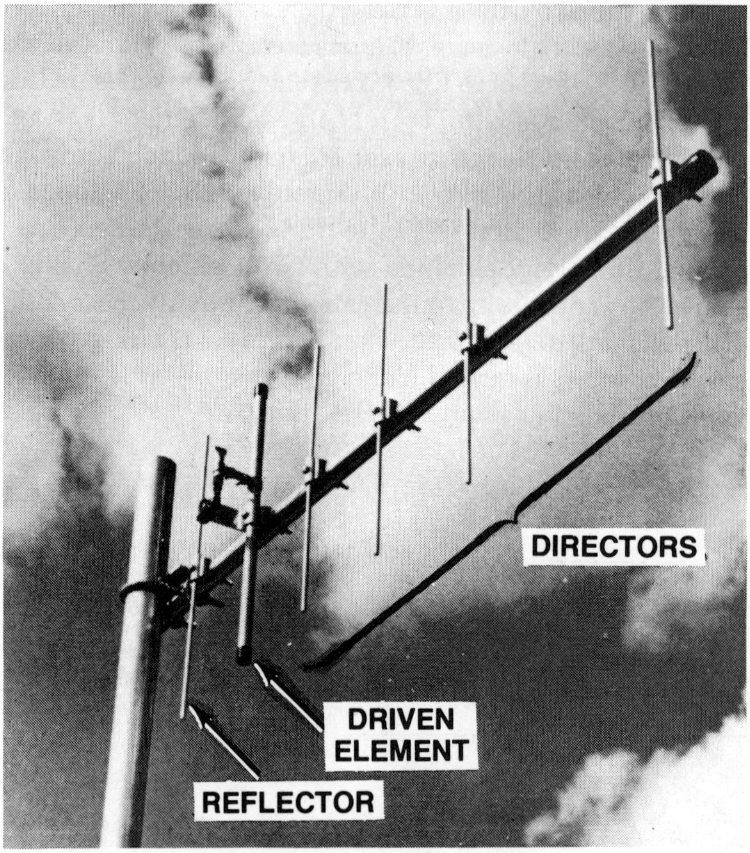

Figure 6-30
A six element Yagi for 440 MHz. Photo courtesy Cushcraft Corp.

DRIVEN ARRAYS

When all the elements of an antenna system are fed by the transmission line, it is called a **driven array**. An example of a driven array is shown in Figure 6-31A. This is called a **collinear array** since all the elements are placed end-to-end or in line. Note that the transmission line is attached to each element. Also, the current in each element is in phase, since the transmission line length to each element is equal. The result is a radiation pattern such as that shown in Figure 6-31B. The dotted line shows the pattern obtained from a single dipole. The dashed line shows the increased directivity obtained with two elements. Finally, the solid line shows the pattern with four elements. Normally these collinear arrays are oriented vertically as pictured in Figure 6-32. This gives omnidirectional coverage but it directs the energy down towards the horizon and thus gives extended coverage at VHF and UHF.

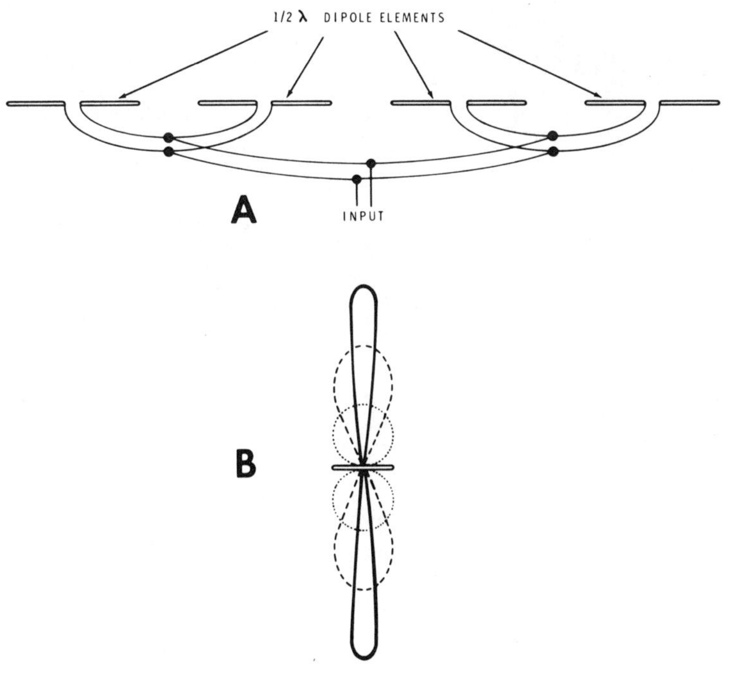

Figure 6-31
A collinear driven array.

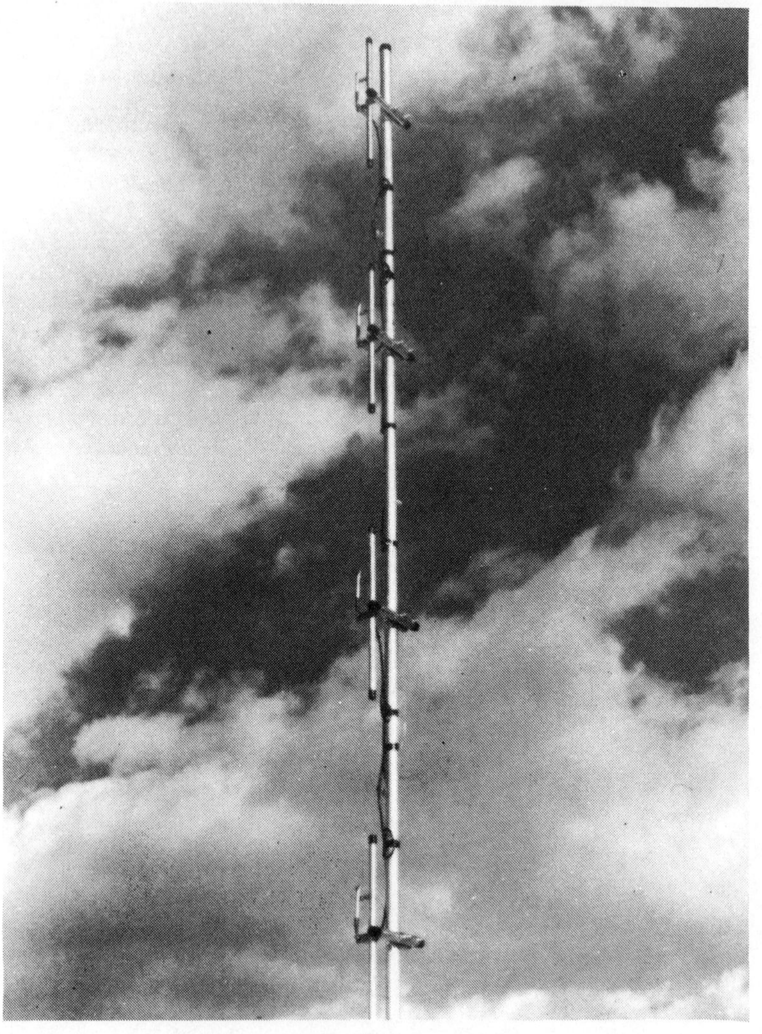

Figure 6-32
A four element collinear array. Photo courtesy Cushcraft Corp.

Another form of driven array is shown in Figure 6-33. This is a **log-periodic array** since the element lengths are related logarithmically. Also, each element is fed through a special phasing network. This antenna offers good directivity and gain. However, its greatest advantage is its very wide bandwidth. While most Yagi arrays operate over a very narrow band of frequencies, the log-periodic array can operate over a much wider range of frequencies. For example, the antenna shown in Figure 6-33 gives high gain and an SWR less than 2:1 from 7.5 MHz to 30 MHz. The best that most Yagi arrays can do is to cover a 1 or 2 MHz band in this same frequency range.

Figure 6-33
A log-periodic array. Photo courtesy Granger Associates.

RADIO PROPAGATION

The energy radiated from a transmitting antenna travels into space in many directions. As the distance from the antenna increases, the energy field spreads out and the field strength decreases. This phenomena was discussed earlier. However, the path, or paths, by which the signal reaches the receiving location also affects the field strength. There are three broad classifications of the signal path. These are: the **ground wave**, the **space wave**, and the **sky wave.**

Ground Waves

The ground wave is a radio wave that travels along the surface of the earth. In the low frequency (LF) and medium frequency (MF) bands, this is the predominant mode of propagation. These longer wavelengths tend to follow the curvature of the earth and actually travel beyond the horizon. This is shown in Figure 6-34. However, as the frequency increases, the ground wave is more effectively absorbed by the irregularities on the earth's surface. This is because, as frequency increases, hills, mountains, etc., become significant relative to the transmitted wavelength. For example, at 30 kHz the wavelength is 10,000 meters or 6.2 miles. Therefore, even mountains are relatively insignificant compared to the wavelength. Thus, very little ground wave attenuation is experienced at this frequency. On the other hand, at 3 MHz the wavelength is 100 meters. This is short enough that hills, trees, and large buildings break up and absorb the ground wave.

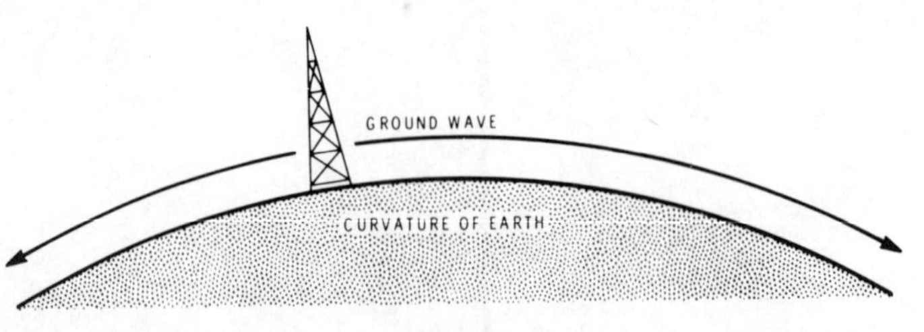

Figure 6-34
The ground wave.

One way of greatly improving ground wave coverage is to use vertical polarization. In fact, with horizontal polarization, the electric field is parallel to the earth's surface, and any ground wave is effectively short-circuited by the conductivity of the earth. However, since ground wave propagation is limited to the low and medium frequency bands, constructing full-size quarter-wave antennas does present problems. For this reason, and others, there are very few communications services in the low frequency band. Most ground wave communications are in the medium frequency bands where the antenna sizes are more practical.

Space Waves

When the transmitted signal is increased above 4 or 5 MHz, the usable ground wave signal exists for only a few miles. At these frequencies and above, particularly at VHF and UHF, it is possible to transmit further using the space wave. When the transmitting and receiving antennas are within sight of each other, the signal is considered a space wave or **direct wave**. This is shown in Figure 6-35. Of course, space wave propagation is limited to line-of-sight distances. Therefore, the range can be increased by increasing the height of both transmitting and receiving antennas. Another form of space wave propagation is transmission to and from satellites and aircraft. Here, the height of the antenna is increased substantially and, thus, the propagation range is greatly increased. Space wave propagation is used primarily in the VHF, UHF, and higher frequency bands.

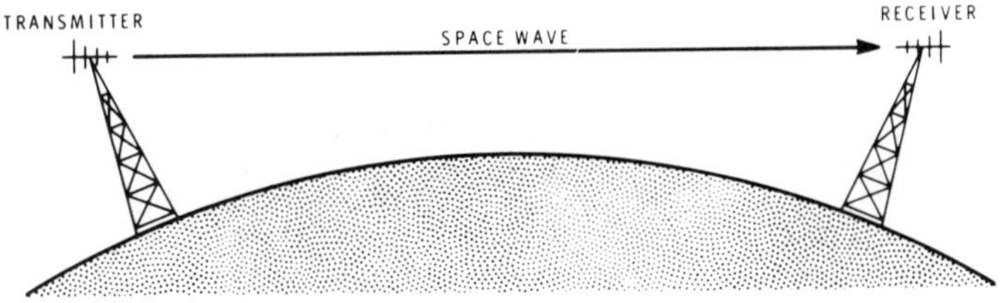

Figure 6-35
Space wave or direct wave propagation.

Sky Waves

From approximately 30 to 250 miles above the earth's surface, ultraviolet radiation from the sun causes the air particles to ionize into free electrons, positive ions, and negative ions. This region is called the **ionosphere**. Any electromagnetic waves that enter this region will be bent or refracted much the same as light waves are bent when traveling through different mediums, such as a prism or lens. The amount of refraction depends on several factors. Among these factors are the frequency of the wave, the density of the ionized region, and the angle at which the wave enters the ionosphere. If all of these factors are just right, the wave will be bent or refracted enough to return to the earth as shown in Figure 6-36. This space wave that has been refracted back to earth by the ionosphere is known as a **sky wave**. Almost all propagation on the high frequency (HF) bands is by sky wave. Also, nighttime long distance medium frequency propagation is by sky wave. Since the ionosphere is responsible for the behavior of the sky wave, it will be discussed in detail.

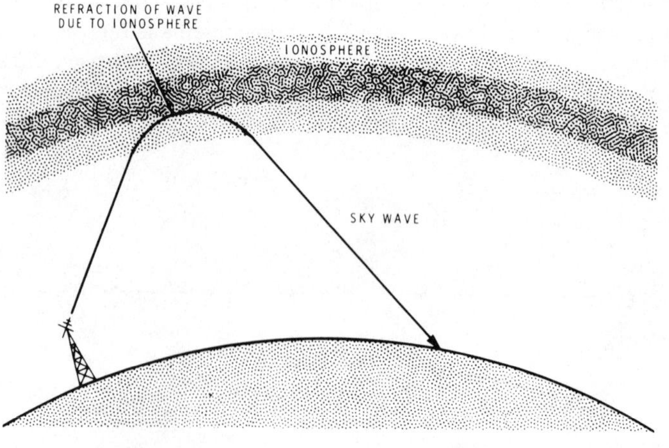

Figure 6-36
The sky wave.

The Ionosphere

Sky waves that return to earth from the ionosphere come from different heights above the earth depending upon frequency and on the time of refraction. This phenomenon shows that several layers of ionization exist. This is because the different gases which make up the earth's atmosphere ionize at different pressures and are also affected differently by ultraviolet radiation and cosmic ray bombardment from the sun. The number of layers present, their heights above the earth and the amount they bend or refract the sky wave all vary, depending upon several interrelated factors. These factors vary from hour to hour, day to day, month to month, season to season, year to year, and even decade to decade. The major influence on the ionosphere is radiation from the sun. For example, solar radiation increases the density of the ionospheric layers and lowers their effective height. This accounts for the hourly and seasonal variations of the ionosphere. Other solar and magnetic disturbances also produce changes in these layers.

Figure 6-37 shows the various ionospheric layers. The lowest layer, called the D layer, exists only in the daytime. This layer is the furthest from the sun and, therefore, its ionization is relatively weak. For this reason, the D layer does not affect the direction of travel of radio waves. However, the ionized particles do absorb appreciable energy from the electromagnetic wave. Thus, when the D layer is present, it attenuates the sky wave on its way up to the other layers and on its way down.

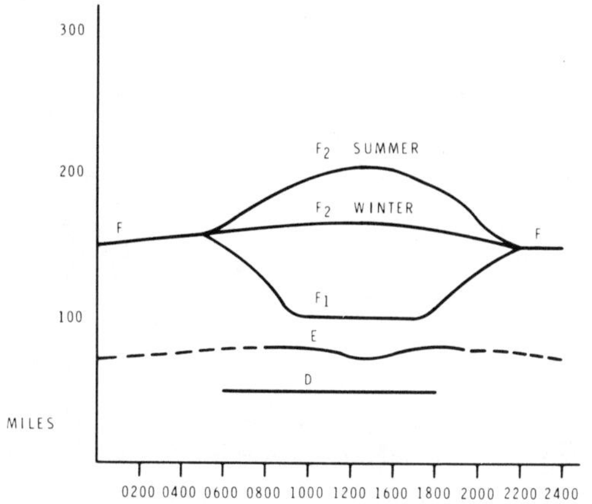

Figure 6-37

The layers of the ionosphere.

The D layer also completely absorbs medium frequency signals. Therefore, it limits MF signals to ground wave propagation during the day. When the D layer disappears at night, MF signals can travel much further via sky wave propagation.

The next layer is the E layer which extends from about 55 to 85 miles above the earth's surface. It has a maximum density at noon, but is only weakly ionized at night, if at all. The last layer is the F layer, which is extremely variable. At night, it is a single layer that is approximately 110 to 250 miles above the earth. However, in the daytime, during maximum radiation from the sun, it splits into two layers called the F_1 and F_2 layers. The F_1 layer ranges from 85 to 155 miles. The F_2 layer, which is the closest to the sun, is even more variable and ranges in height from 90 to 185 miles on a winter day, to as high as 155 to 220 miles on a summer day.

Sky Wave Propagation

This section lists and defines several terms that are used to describe characteristics of the ionosphere and sky wave propagation. The first of these terms is virtual height. Figure 6-38 shows that as the wave is refracted, it is bent down gradually rather than sharply. However, below the ionized layer, the waves follow paths that are exactly the same as if they were "reflected" from a surface located at a greater height, called the **virtual height** of the layer.

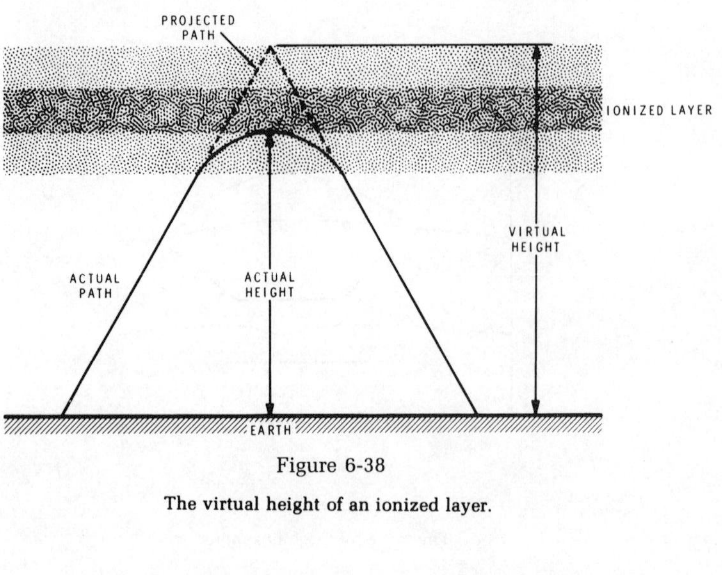

Figure 6-38

The virtual height of an ionized layer.

The virtual height of the ionospheric layers can be calculated from the time it takes a pulse of RF energy, directed vertically upward, to be returned to earth. However, if the signal frequency is raised above a certain value, it will not be refracted enough to return to earth. This **critical frequency** is defined as the highest frequency that is returned to earth when transmitted in a vertical direction. The critical frequency value depends on the condition of the ionosphere and, therefore, changes from hour to hour, day to day, etc.

If the angle of radiation is lowered from vertical, the wave will travel longer through the ionized layer and will, therefore, be refracted to a greater degree. This means that signals above the critical frequency can be returned to earth. However, there is a limitation. For a given frequency there is a **critical angle** beyond which the signal will not be refracted enough to return to earth. This is shown in Figure 6-39. Note also that as the radiation angle decreases, the distance the wave travels over the earth increases. This distance is known as the **skip distance**. The skip distance can be maximized by using the lowest radiation angle possible and by using the highest frequency that will be refracted at that angle.

The highest frequency that can be used for transmission between two locations is called the **maximum usable frequency** (MUF). Actually, a lower frequency could be used, as it would also be refracted. However, as the signal frequency is lowered, the energy absorbed in the ionosphere increases rapidly, and the signal level is drastically reduced. In fact, there is a lowest usable frequency (LUF), below which the RF signal is absorbed entirely by the ionosphere. Therefore, maximum received signal level is obtained when operating at the maximum usable frequency.

Unfortunately, the maximum usable frequency is constantly changing due to the effects of solar radiation. Therefore, for the most reliable sky wave communications, the **optimum usable frequency** is used. This frequency is far enough below the MUF that it isn't affected by minute-to-minute solar fluctuations and, thus, provides reliable sky wave communications. Since MUF, LUF, and optimum frequencies change hourly, daily, and monthly, it may seem that sky wave communications is a hit-and-miss proposition. However, there are charts available that predict the MUF, LUF, and optimum frequencies for every hour of the day, over any path on earth, during a given month. These predictions are based on elaborate solar observations and can be used to optimize sky wave communications.

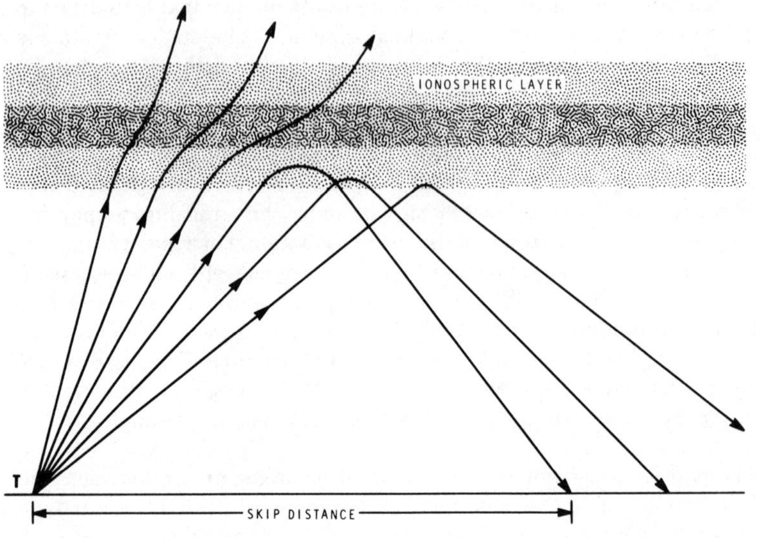

Figure 6-39
Effects of lowering the radiation angle.

MULTIPLE-HOP TRANSMISSION

Figure 6-40 shows that if the signal that has returned to earth at location A has sufficient strength, it can be reflected by the ground back up to the ionosphere. It is then refracted again and it will return to earth at location B. Here again, if it has sufficient strength, it will be reflected again and continue to a more distant location. This is known as multiple-hop transmission and it is by no means limited to two or three hops.

Under optimum conditions, the maximum distance that a single-hop transmission can cover is about 2000 miles. The limiting factor is the radiation angle which, of course, cannot be reduced below the horizon. Therefore, for long distance coverage multiple-hop propagation is necessary. However, remember that each hop increases the attenuation of the signal. Thus, attenuation from the ionosphere and the ground reflection points is the primary factor in determining just how far a sky wave will travel.

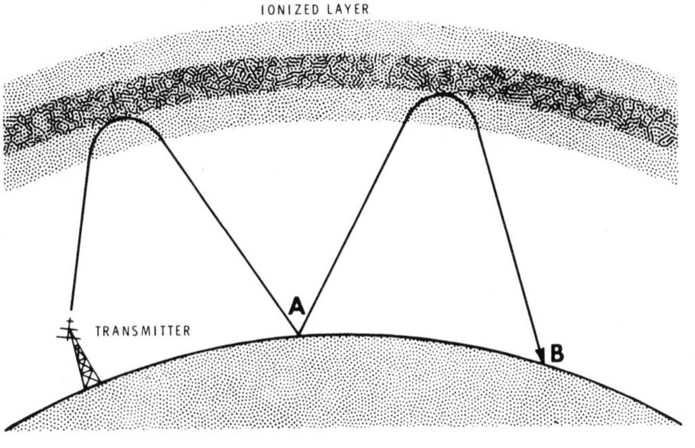

Figure 6-40

Double-hop transmission.

FADING

When sky wave propagation is used, signal strength will increase and decrease periodically. Sometimes, the changes in signal strength are small and the receiver's automatic gain control (AGC) will compensate for it. Other times, the signal may be lost completely. This is known as **fading** and is caused by multiple-path reception and changing ionospheric conditions. For example, consider the condition shown in Figure 6-41. Here, the signal is received via both one and two hop paths. In this case, the lengths of the paths and the phasing of the signals will vary with ionospheric conditions. Because the signal received at any instant is the phasor sum of all the waves received, alternate cancellation and reinforcement will occur if the path lengths vary over 1/2 wavelength. Thus, changing ionospheric conditions which affect path lengths will cause fading.

It is also possible to receive the same signal via both sky wave and ground wave propagation. If they do not arrive in the correct phase, fading will result. This occurs primarily in the medium frequency band.

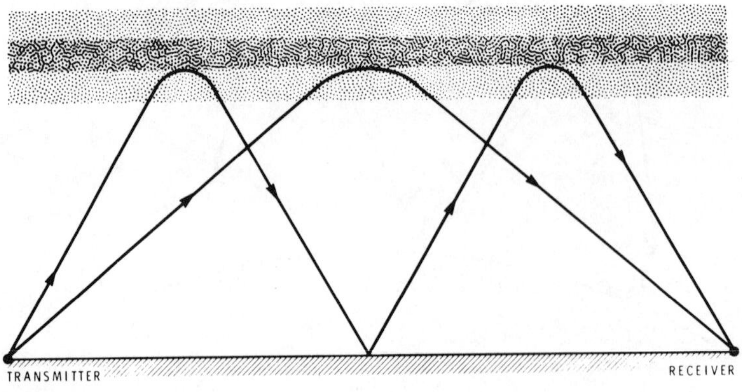

Figure 6-41
Multiple path reception.

Ionospheric Variations

Since the existence of the ionosphere depends on solar radiation, it follows that any variations in this radiation will influence the ionosphere. The earth's rotation and its revolution around the sun also affect the amount of solar radiation reaching the earth and therefore, they too influence the ionosphere.

The regular variations of the sun, earth, and hence the ionosphere which are more or less predictable are divided into the following categories: diurnal, seasonal, geographical, and cyclical. Diurnal variations are the hour-to-hour changes in the various layers caused by the rotation of the earth around its axis. Diurnal variations in the D, E, and F_1 layers follow a regular pattern and are dependent on the sun's elevation above the horizon. Ionization in these layers increases from a very low level at sunrise, to a maximum at noon, and then decrease toward sunset. For all practical purposes, these layers disappear at night. Typical hour-to-hour changes in the reflection capability of the various layers, as indicated by the critical frequency, are shown in Figure 6-42. Notice that ionization in the F_2 layer rises steeply at sunrise. It decreases after sunset, but does not disappear during the night.

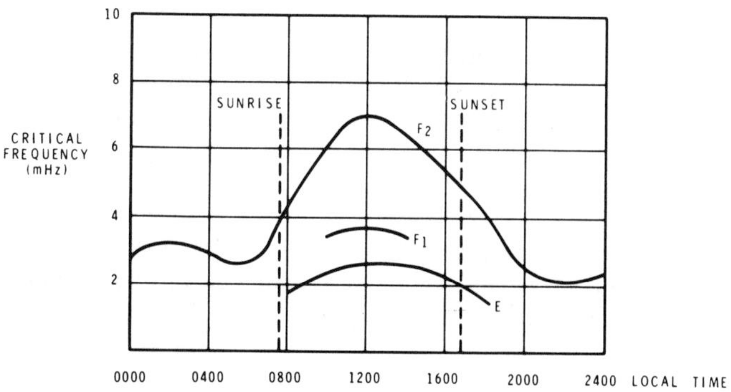

Figure 6-42

Diurnal variations of the ionosphere.

Since the position of any point on earth relative to the sun is constantly changing as the earth moves in its year-long orbit around the sun, ionospheric properties change also. These are called seasonal variations, and a typical variation in the F_2 layer is shown in Figure 6-43. Note that since the earth is closer to the sun in winter, it receives a higher level of radiation, and the critical frequency reaches its highest level.

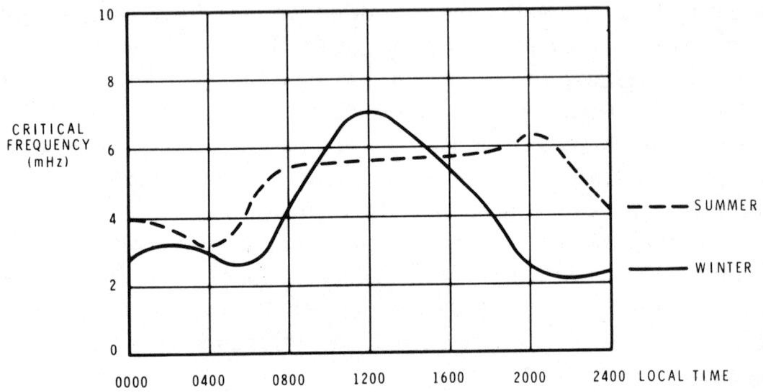

Figure 6-43

Seasonal variations of the F_2 layer.

The intensity of ionizing radiation that strikes the ionosphere also varies with latitude. For example, solar radiation is considerably greater near the equator, where the sun is more directly overhead, then in the northern and southern latitudes. These variations in the ionosphere are called geographical variations.

The 11-year sunspot cycle probably influences the ionosphere most. This variation depends upon sunspot activity, which is constantly changing throughout an approximate 11-year cycle.

A

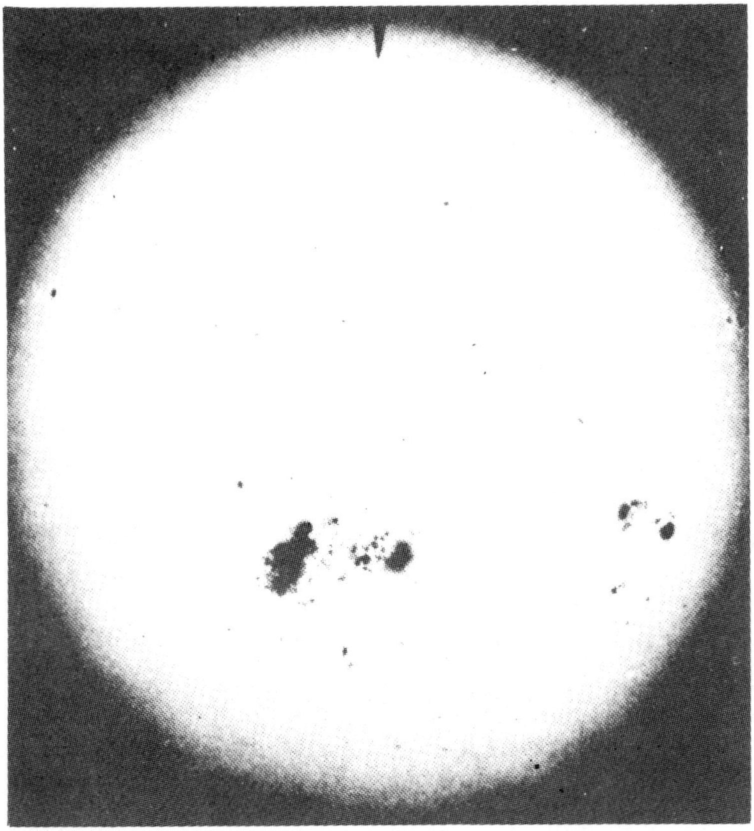

B

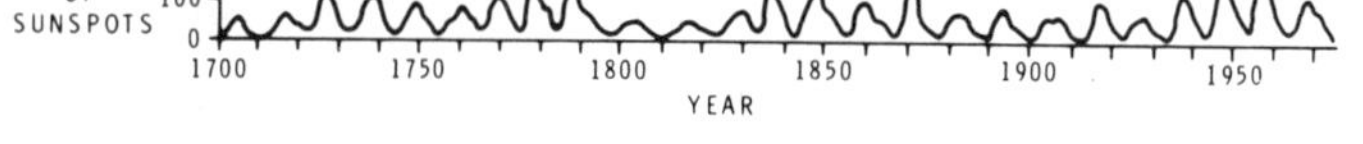

Figure 6-44

Sunspots greatly affect the ionosphere.
(A) shows the surface of the sun with a number of sunspots.
(B) shows the cyclical variation in the sunspot count.

A photograph of the surface of the sun that shows a number of sunspots is shown in Figure 6-44A. A direct relationship exists between these sunspots and the intensity of ultraviolet energy radiated by the sun: the greater the number of sunspots, the greater the intensity. Therefore, since ultraviolet radiation is responsible for ionizing the various layers, it follows that the sunspot count directly affects the condition of the ionosphere. This effect is dramatically illustrated in Figure 6-45. It shows a comparison between critical frequencies for the F_2 layer during a sunspot maximum and a sunspot minimum. Notice that the critical frequency during a sunspot maximum is over twice that of a sunspot minimum. The result is that communications range and reliability is greatly increased during years of high sunspot activity.

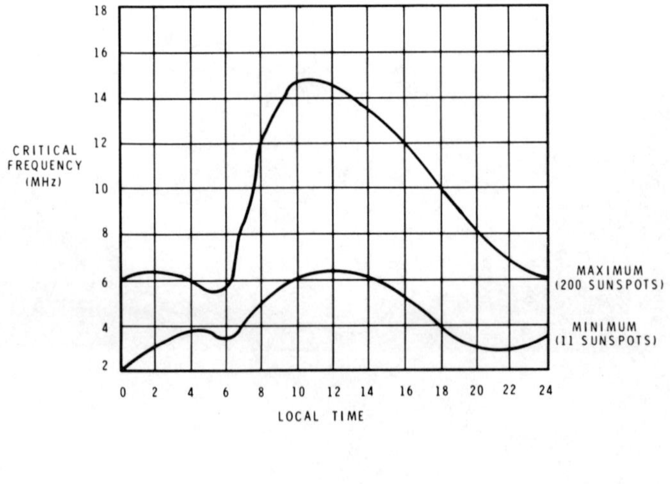

Figure 6-45

Comparison of the critical frequency of the F_2 layer versus the number of sunspots.

The ionosphere is also subject to abnormal variations which can cause unpredictable changes in wave propagation. Generally, ionospheric disturbances on the HF bands weaken signal levels, either suddenly or gradually, sometimes to the point where signals disappear completely. On the other hand, in the VHF bands, these ionospheric disturbances can produce very long distance communications "openings". One of these disturbances is known as "sporadic E-layer ionizations". These are erratic areas of high density ionization that occur at E-layer heights. This usually results in ionospheric refraction of signals high up into the VHF region, and it allows extremely distant contacts on these normally line-of-sight frequencies.

The other two main types of ionospheric disturbances are the "ionospheric storm" and the sudden ionospheric disturbance. In both cases, they are believed to have their origin in solar flares which occur on the surface of the sun. An ionospheric storm may develop either gradually or suddenly, and may continue from one or two days to almost a week. The sudden ionospheric disturbance, on the other hand, usually commences suddenly and lasts from about twenty minutes to an hour or two. At the height of these disturbances, the absorption is high enough to cause a radio blackout, particularly at the higher frequencies.

VHF-UHF Propagation

The maximum usable frequency seldom rises above 30 MHz and only at the sunspot maximum does it reach as high as 50 — 60 MHz. For this reason, most communication in the VHF and UHF bands takes place via the space wave and is, therefore, strictly limited to line-of-sight paths. Thus, the limiting factor for VHF-UHF communications is the local terrain and the curvature of the earth.

Actually, the **radio horizon** for space waves is about 1-1/3 as far as the optical horizon. This is due to refraction in the earth's lower atmosphere. Refraction occurs because of the linear decrease in density of the earth's atmosphere as height increases. What happens, in effect, is that the top of the wave travels slightly faster than the bottom. As a result, the wave is bent slightly downward to follow the earth's curvature beyond the optical horizon.

The radio horizon of an antenna is given by the formula:

$$D_t = 4\sqrt{H_t}$$

where D_t = radio horizon distance in km.

H_t = transmitting antenna height in m.

You can use the same formula for a receiving antenna. Therefore, the maximum space wave communications distance is the sum of these figures or

$$D_m = D_t + D_r = 4\sqrt{H_t} + 4\sqrt{H_r}$$

As an example, consider a transmitting antenna height of 100 m and a receiving antenna height of 49 m. What is the maximum space wave communications distance?

$$D_m = 4\sqrt{H_t} + 4\sqrt{H_r}$$

$$= 4\sqrt{100m} + 4\sqrt{49m}$$

$$= 4(10) + 4(7)$$

$$= 40 + 28$$

$$= 68 \text{ km}$$

In this case, the maximum communications distance is 68 km. Of course, this does not account for any path obstructions, such as mountains, hills, buildings, etc. All these factors must be considered when setting up a communications link.

Extending VHF-UHF Communications

There are many ways of extending the range of distance of VHF-UHF and higher frequency communications. However, only two techniques have proven reliable enough to merit widespread use. These are tropospheric scatter communications and satellite communications.

TROPOSPHERIC SCATTER PROPAGATION

Figure 6-46 illustrates a typical **troposcatter** communications link. Here, a very high power UHF transmitter uses a high gain directional antenna to transmit energy at the horizon. This energy is then "scattered," with some minute portion being scattered in the direction of the receiver. The reasons for the scattering are not entirely understood. However, one theory suggests that the signal is reflected from irregularities in the atmosphere, another suggests reflection from atmospheric layers. At any rate, the phenomena does exist and is a quite reliable method of extending UHF communications beyond the horizon.

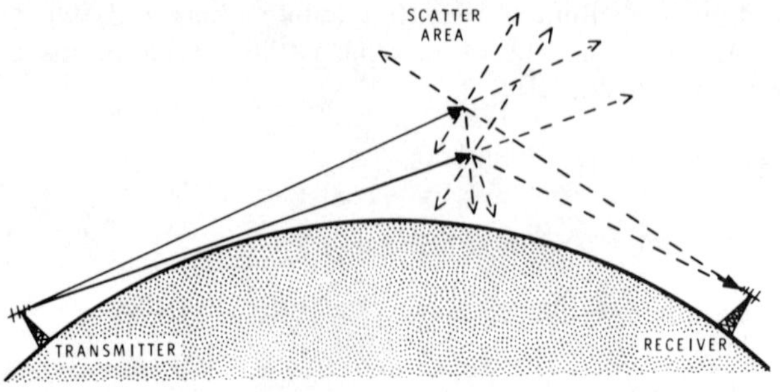

Figure 6-46
Tropospheric scatter propagation.

The best frequencies for this mode are 900 MHz, 2 GHz, and 5 GHz. The typical troposcatter path length is 300 to 1000 km. Compared to the maximum practical space wave distance of 100 km, troposcatter appears quite attractive. The only problem is that a high power transmitter, a high gain receiver, and elaborate antenna arrays must be used. Therefore, the cost of a troposcatter system is relatively high compared to using several space wave repeater stations or a coaxial cable to cover the same distance. Even so, troposcatter systems are used when the communications path must cross rough or inaccessible terrain. As an example, they are used almost exclusively to provide communications to offshore oil drilling platforms in the North Sea and elsewhere.

SATELLITE COMMUNICATIONS

A communications satellite is basically a microwave repeater station. It receives an "uplink" signal from earth, amplifies it, and retransmits it on a separate "downlink" frequency. By virtue of its position above the earth, it allows any station in its coverage area enormously increased communications range. Figure 6-47 shows a typical satellite communications system. Most systems presently use a 6 GHz uplink frequency and a 4 GHz downlink frequency. In addition, there are several separate communications channels in each satellite. In fact, presently operational satellites can provide up to 11,000 telephone circuits each!

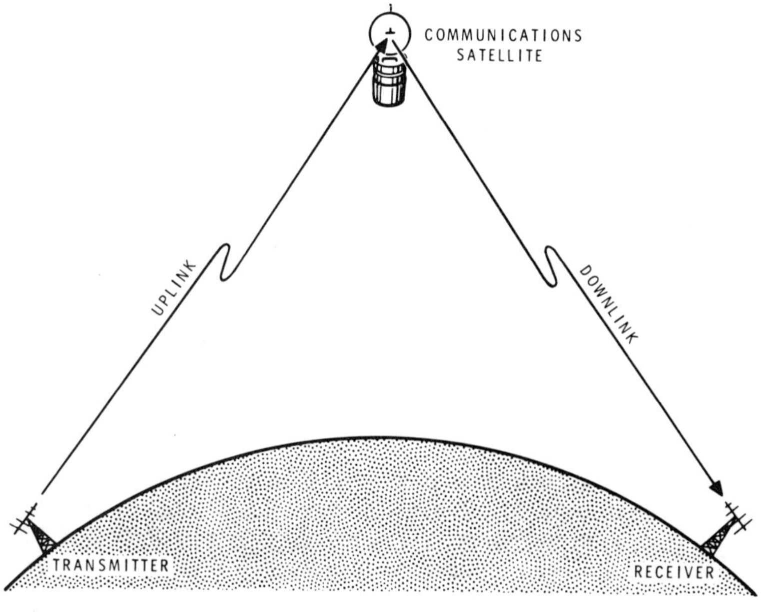

Figure 6-47
A satellite communications system.

The vast majority of communications satellites now in use are "parked" in a synchronous orbit. That is, they appear to hover over a certain spot above the earth. To do this, they must be precisely 35,800 km above the earth's equator. Once this is accomplished they will complete one orbit every 24 hours, just as the earth completes one revolution. Therefore, they will have the same angular velocity as the earth and will remain above the same spot on earth. Thus, the satellite is **geostationary**, or stationary with respect to the earth.

Once a satellite is parked in its synchronous orbit, it is a relatively simple matter to aim the ground station antennas. This is because the satellite is always at a precise point above the horizon. This eliminates costly satellite tracking equipment that is required with satellites in other types of orbits.

Figure 6-48 shows the coverage pattern of a geostationary satellite using an omnidirectional antenna. The satellite is parked at approximately 15° west longitude directly above the equator. From this map, you can see that it would be a relatively simple matter to cover the entire earth, excluding the polar regions, using just three equally spaced satellites. In practice, many more are used. Some are used simply to provide more channels or communication capacity. Others provide more specialized applications or coverage. For example, the ANIK series of satellites is used to provide communications within Canada. As a result, they can use directional antennas on the satellite to provide better coverage of Canada and, in particular, its far northern regions.

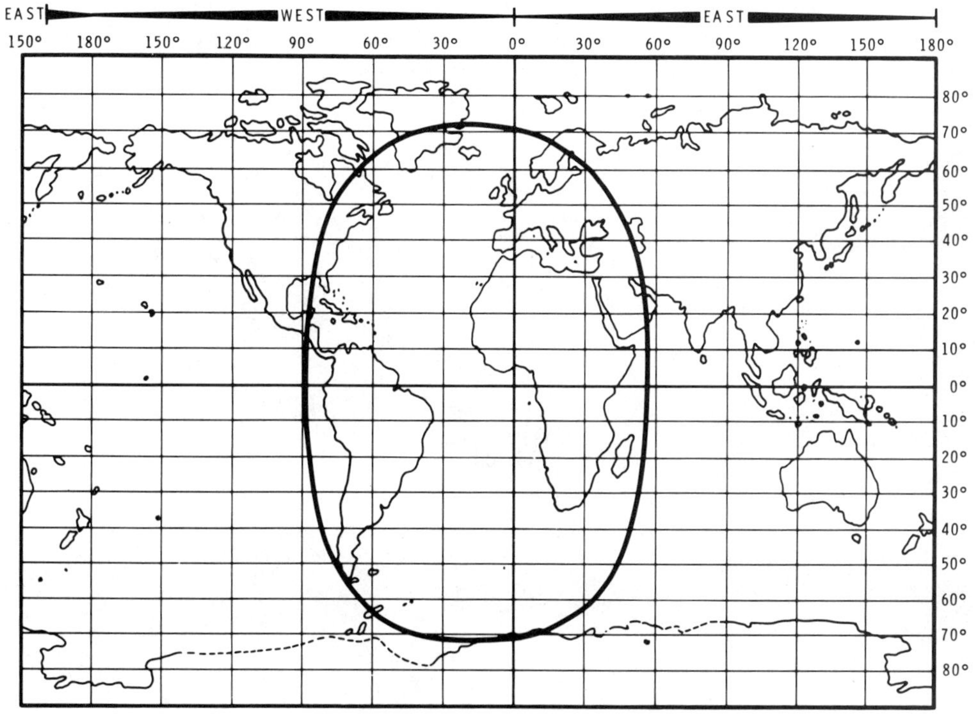

Figure 6-48

Coverage of a geostationary satellite parked at 15° W.

An example of a domestic satellite is shown in Figure 6-49. This satellite uses a highly directional antenna to provide coverage of the continental United States. The satellite consists of the directional antenna (which is used for both transmit and receive) and a receiver/transmitter which is called a transponder. The receiver does not demodulate the signal, instead, it amplifies the 6 GHz uplink signal and heterodynes it down to 4 GHz. The transmitter then amplifies this 4 GHz downlink signal and applies it to the antenna. The overall amplification is approximately 130 dB and the total power consumption about 150 W. The power is provided by batteries which are recharged by solar cells.

Since the satellite's power consumption must be kept low, the transmitter's output power is also kept low. This means that the earth stations must use highly directional antennas, high transmitter power, and very high gain receivers. However, this is more than offset by the increased communications range and reliability.

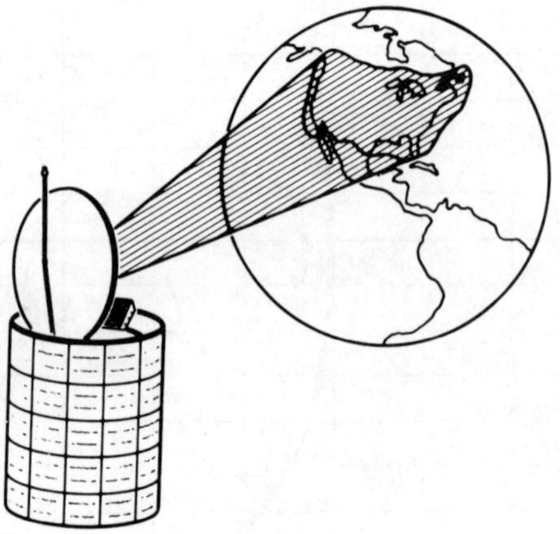

Figure 6-49.
A domestic satellite and its coverage area.

Unit 7

COMMUNICATIONS SYSTEMS

INTRODUCTION

This unit discusses three basic communications systems. These are the television broadcasting system, the FM stereo broadcasting system, and several methods of data communication.

FM STEREO BROADCASTING

Stereo (or stereoscope) originally referred to a special photographic technique used to give the viewer the impression of observing a scene in three dimensions. This was accomplished using two photographs of the same object taken from slightly different angles. When one photo was viewed by one eye and the other photo by the other eye, the result was the appearance of a three-dimensional image. Almost exactly the same technique is used in stereophonic audio. Here, the same sound source (speaker, orchestra, etc) is recorded from two different angles, in this case, the left and right sides. When the recording is "played back" through left and right loudspeakers to the left and right ears, the result is the appearance of a "three-dimensional" sound source. It lends realism to the recorded sound just as the stereoscope brought the two-dimensional photograph to life.

Adapting stereo audio to broadcasting presented two problems. First, any system used to transmit stereo signals had to be compatible with the existing monophonic receivers. And second, the stereo signal had to be transmitted within the 200 kHz bandwidth allotted to each FM station. Eventually, a system was approved that used a frequency division multiplex scheme.

The Stereo Multiplex System

To provide the necessary compatibility with monophonic receivers, the stereo multiplex system transmits both the right and left channels at their normal audio frequencies from 50 Hz to 15 kHz. This is called the **L+R** signal. The system also transmits an **L−R** signal. It is transmitted as a double-sideband suppressed-carrier signal on a 38 kHz subcarrier. As you learned earlier, a double-sideband suppressed-carrier signal is extremely difficult to demodulate without a synchronizing signal. For this reason, a special **pilot carrier** is transmitted at 19 kHz, exactly one-half the frequency of the 38 kHz sub-carrier. The resulting stereo multiplex signal spectrum is shown in Figure 7-1. This multiplex signal is then used to modulate the FM broadcast transmitter.

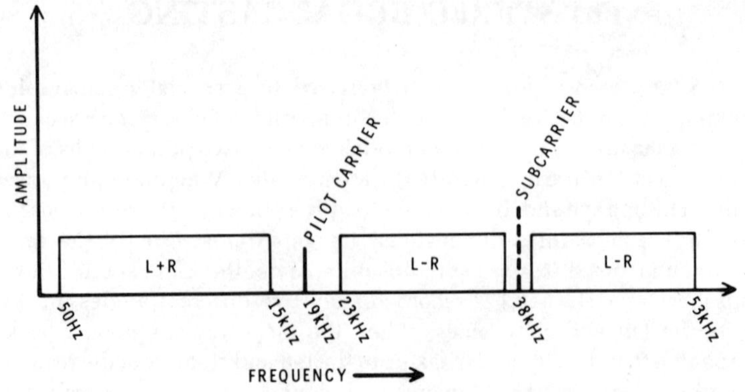

Figure 7-1
The Stereo Multiplex Spectrum.

Once the stereo multiplexed FM signal is received and demodulated, its spectrum again appears just as shown in Figure 7-1. A monophonic receiver, whose audio response is limited to 16 or 17 kHz, utilizes only the L+R signal. Therefore, it reproduces both audio channels, but without any stereo separation. Nonetheless, the multiplex system does produce a compatible monophonic signal containing both left and right channels. On the other hand, a stereophonic receiver uses special demultiplexing circuits to obtain separate left and right audio channels. Before examining the stereo receiver in detail, let's look at the stereo transmitter.

The Stereo Transmitter

The block diagram of a typical stereo multiplex transmitter is shown in Figure 7-2. There is a microphone and a preamplifier for both the left and right channels. Both the left and right signals are immediately applied to summing amplifier "A." Its output is a linear mix of the two channels and is, therefore, the L+R signal.

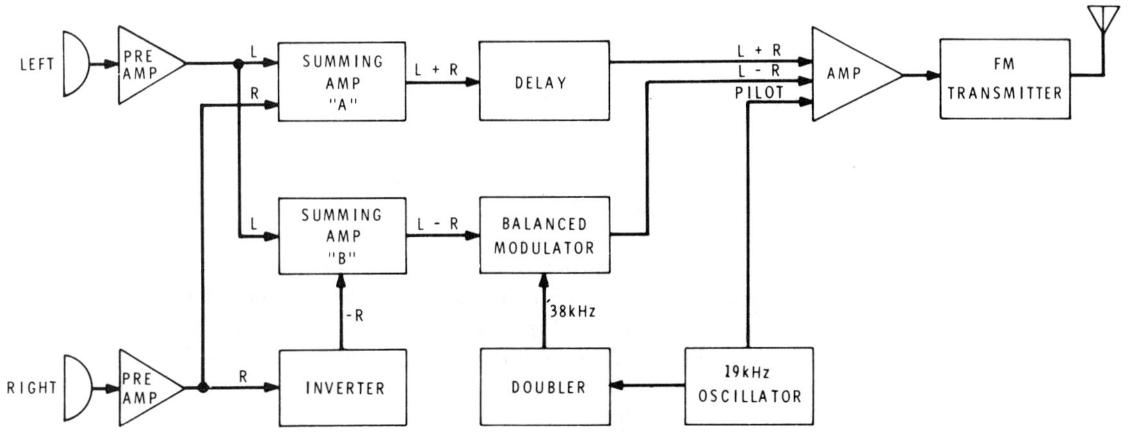

Figure 7-2
A Stereo Multiplex Transmitter.

To form the L−R signal, the left channel is applied directly to summing amplifier "B". However, the right channel passes through an inverter to become the −R signal before being applied to the summing amplifier. Therefore, the output of summing amplifier "B" is a linear mix of the left channel and an inverted right channel, or the L−R signal.

The L−R signal is then applied to a balanced modulator. A 38 kHz carrier is also applied. This carrier is derived from a 19 kHz master oscillator using a frequency doubler. The output of the balanced modulator is the double-sideband suppressed-carrier L−R signal. This signal is then coupled to the FM transmitter through an amplifier. It is joined by the L+R signal which passed through a delay network to compensate for balanced modulator delay of the L−R signal. The 19 kHz pilot is also applied to the transmitter.

In FM broadcasting, the maximum allowable carrier deviation is ±75 kHz. This is true for either monaural or stereo programming. Therefore, when the modulating signal is stereo multiplex, deviation by the program audio is slightly reduced. This is due to the 19 kHz pilot carrier. It is always present and does not vary in amplitude. It deviates the carrier 10% of its maximum deviation or ±75 kHz × 10% = ±7.5 kHz. Thus, the maximum deviation for the audio signals is 90% of ±75 kHz or ±67.5 kHz. The difference between 90% and 100% deviation is minimal and not readily discernible to the listener. It may also seem that this remaining 90% deviation must be divided equally between the L+R and L−R signals. However, in practice, a process called **interleaving** takes place. When the L+R signal is at its peak amplitude, the L−R signal is at its minimum amplitude, and vice versa. Therefore, due to interleaving, each modulating signal can independently produce almost full deviation without over deviating the carrier.

The Stereo Receiver

One example of a stereo multiplex receiver is shown in Figure 7-3. A normal FM tuner and demodulator are used to recover the actual stereo multiplex signal. Then, filters are used to separate the basic components of the signal. A low pass filter is used to obtain the L+R signal, which is 50 Hz to 15 kHz. A bandpass filter separates the 23-53 kHz double-sideband L−R signal. And finally, a 19 kHz bandpass filter separates the pilot carrier.

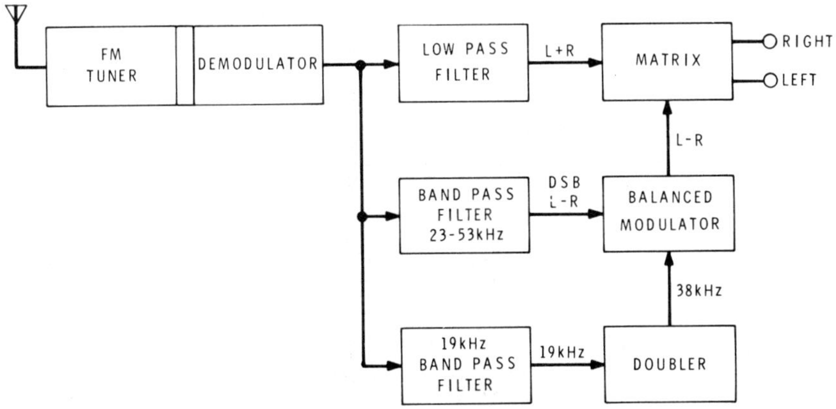

Figure 7-3
A Stereo Multiplex Receiver.

The 19 kHz pilot is then applied to a doubler to obtain the correct phase and frequency (38 kHz) subcarrier. This is coupled to the balanced modulator along with the double-sideband L−R signal. The demodulated L−R signal and the L+R signal are then combined in a special matrix network to obtain the separate left and right channels.

Figure 7-4 shows the schematic diagram for a very basic matrix network. Here, the 38 kHz subcarrier has been developed and is applied to the junction of D_1D_2. The FM demodulator output is also applied. The L+R signal is separated by the low pass filter of C_1L_1 and developed across R_1. The double-sideband L−R signal is applied to the junction of D_1D_2.

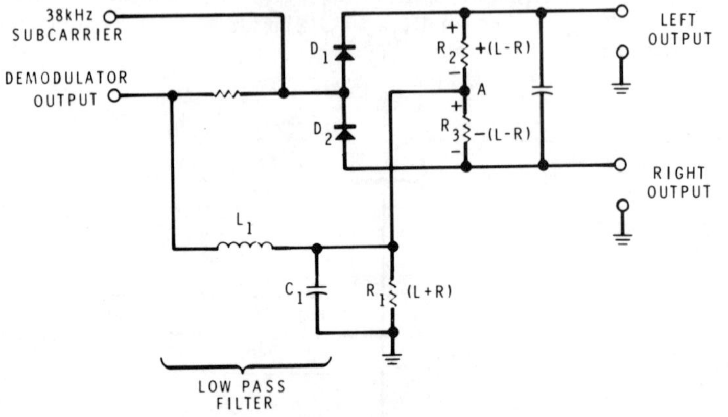

Figure 7-4
Simplified Stereo Multiplex Matrix.

The L−R signal is demodulated by D_1D_2 and developed across R_2 and R_3. However, due to the diodes' polarity, the signal across R_2 is 180° out of phase with the signal across R_3, with respect to point A. Note that the L+R signal is added at point A. This means that the left output "sees" a +(L−R) signal across R_2 and the (L+R) signal across R_1. These two signals add to produce the left output, because (L+R) + (L−R) = 2L. Similarly, the right output "sees" a − (L−R) signal across R_3 and the (L+R) signal across R_1. Therefore, it produces a right output because (L+R) − (L−R) = L + R − L + R = 2R.

Stereophonic sound results when these two signals are amplified and fed to loudspeakers. The examples shown were simplifications and you should remember that many circuit variations exist. For example, the FM tuner of schematic diagram #2, the Heathkit AJ-1219, has an integrated circuit stereo demodulator.

SCA

There is one other signal that can be multiplexed onto an FM broadcast transmission. It is an SCA or Subsidary Communication Authorization. The SCA that most of us are familiar with is Muzak® or storecasting. This is a special background music service that contains no advertising or announcements. It is paid for by subscription rather than advertising revenue; and is transmitted along with the normal FM broadcast, except that it is multiplexed onto a special subcarrier. This subcarrier can be any frequency from 20 kHz to 75 kHz. However, if the station is also transmitting a stereo multiplex signal, the SCA subcarrier must be above the L−R signal, or from 53 kHz to 75 kHz. The accepted standard SCA frequency is 67 kHz.

The SCA subcarrier is frequency modulated with a maximum deviation of ±7.5 kHz. The spectrum of a stereo generator output with an SCA subcarrier is shown in Figure 7-5. This entire spectrum is then used to modulate the FM broadcast transmitter. The SCA signal is only allowed to deviate the FM transmitter 10% of its maximum of ±75 kHz. Therefore, since the stereo pilot also deviates the transmitter 10% of maximum, this leaves only 80% of deviation for the L+R and L−R signals. Thus, the maximum allowable deviation for the main channel is reduced by 10% when adding an SCA signal.

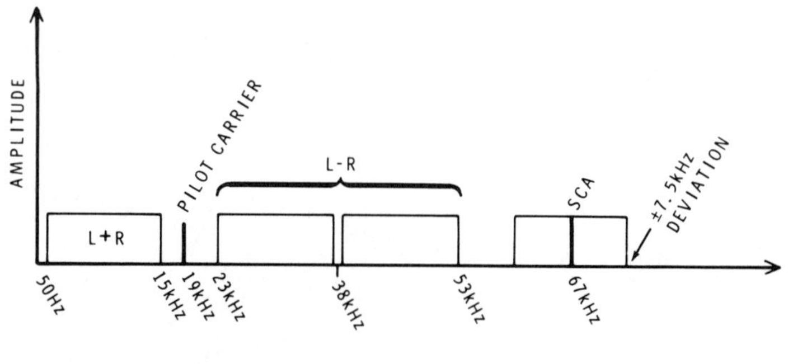

Figure 7-5
Spectrum of a Stereo Multiplex Generator with an SCA Signal.

®Trademark, Muzak Corporation.

SCA program material can be any number of things. Usually it is continuous background music as mentioned previously. However, it can be special purpose programming for particular audiences such as "talking books" for the blind or even slow scan television for the deaf. One application is relaying telemetry signals to the studio from a remote transmitter site.

The SCA receiver is relatively simple. After a normal FM tuner's demodulator, a 67 kHz (or any other SCA frequency) bandpass filter is used to separate the SCA signal. Then a phase-locked loop tuned to 67 kHz is used to demodulate the FM signal. The detected signal is amplified and fed to a loudspeaker or other output device. The listener can use the main channel program, the SCA program, or both, simultaneously.

Pre-emphasis and De-emphasis

In FM broadcasting, a method of noise reduction is used that requires pre-emphasis at the transmitter and de-emphasis at the receiver. Pre-emphasis means that the higher audio frequencies are increased in amplitude. De-emphasis means that the higher audio frequencies are decreased in amplitude. The reason for this is to overcome the noise level and thereby increase the system's signal-to-noise ratio.

Noise is always present in any system. However, as the audio frequency increases, so does the noise level. Therefore, if the audio signal is increased proportionately to the noise level, the signal-to-noise ratio will improve greatly at the higher frequencies.

In FM broadcasting, the program audio signal is fed through a special RLC pre-emphasis network whose time constant is 75 μs. A graph of the output signal amplitude versus frequency is shown in Figure 7-6. This output response curve is very similar to a high pass filter response. To stay within FCC regulations, the transmitted audio signal must be within the solid and dashed lines shown on the chart.

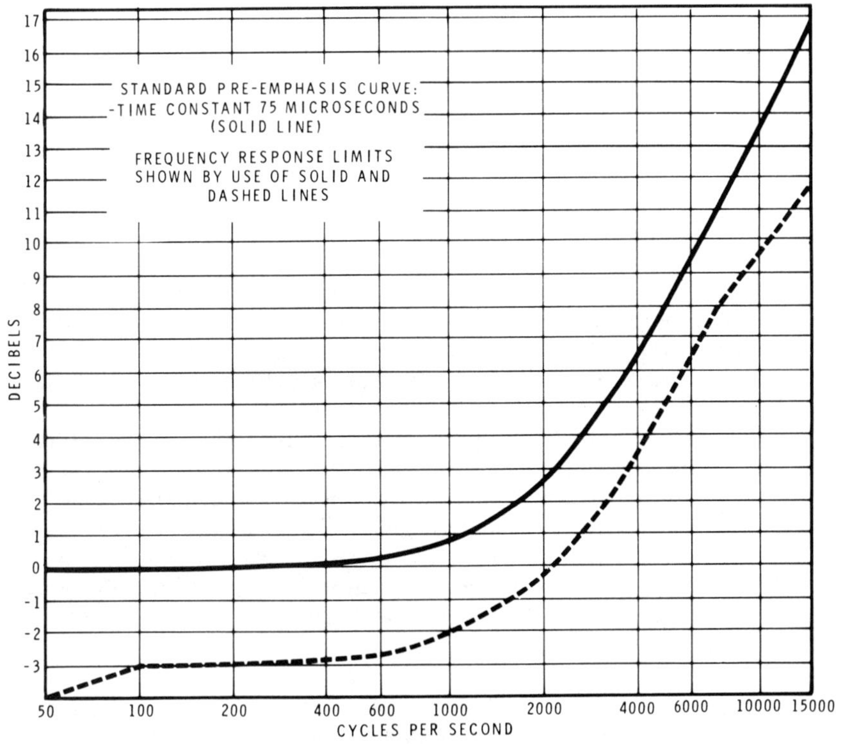

Figure 7-6
Standard 75 μs Pre-Emphasis Curve.

At the receiver, an RC or RLC de-emphasis network is used at the audio amplifier. Its response curve should be exactly opposite the 75 μs pre-emphasis curve. This restores the audio signal to its original frequency -versus- amplitude response. Thus, the audio frequency response is returned to normal, while the signal-to-noise ratio is greatly improved.

TELEVISION BROADCASTING

Of the three basic communications systems discussed in this unit, television broadcasting is perhaps the most widely used and by far the most popular form of entertainment. One reason for this may be that it stimulates not only hearing, as does FM broadcasting, but also sight. This allows much greater creative freedom in the programming content.

The TV broadcasting system consists of both audio and video pickup devices, a transmitter, and a receiver. This section covers all of these important components.

Television Transmission

Every television broadcast consists of two separate signals: an audio signal and a video signal. The audio signal is broadcast by an FM transmitter whose maximum deviation is ±25 kHz. Other than the lower deviation, it is virtually identical to a monaural FM broadcast transmitter.

The video signal is the output of a special camera tube, such as a vidicon or plumbicon, that converts light into an electrical signal. A microphone does much the same thing for audio by coverting sound waves into electrical signals. However, in most visual scenes, light levels vary much more rapidly than do the sound waves in music or speech. For this reason, the video signal's frequency can be as high as 8 MHz, although for TV broadcasting it is limited to 4.2 MHz.

The video signal is used to modulate a separate AM transmitter. However, since the bandwidth would be 2 × 4.2 MHz or 8.4 MHz, a compromise system of attenuating the lower sideband is used to reduce bandwidth. This is called **vestigial sideband transmission**, since only a vestige, or small portion, of the lower sideband is transmitted.

The frequency spectrum of a television transmission is shown in Figure 7-7. It shows that the vestigial sideband extends 1.25 MHz below the carrier. The majority of visual information is contained within this 1.25 MHz band. Therefore, both sidebands transmit this important part of the video spectrum, with the full upper sideband transmitting the finer visual detail contained within the band from 1.25 MHz to 4.2 MHz. The aural carrier with its ±25 kHz deviation is positioned exactly 4.5 MHz above the visual carrier. Thus, the entire television broadcast signal occupies a 6 MHz bandwidth. A further bandwidth reduction could be obtained by using a single-sideband visual signal. However, the required receiver would be extremely complex and expensive. Thus, the compromise vestigial sideband system was chosen.

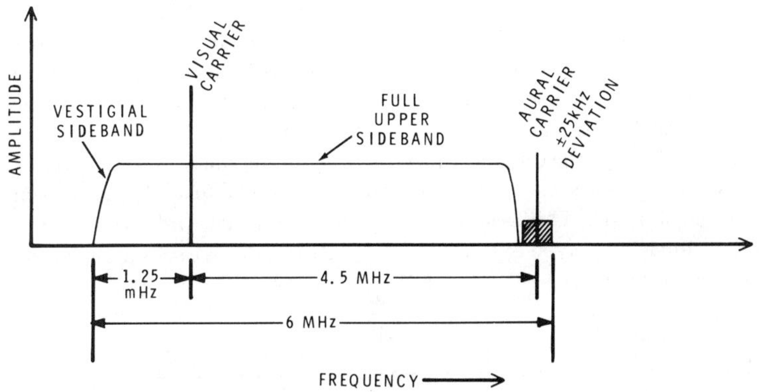

Figure 7-7

Television Transmission Spectrum.

Television transmissions are separated into three frequency bands. These are:

Channels 2 through 6 — 54 to 88 MHz
Channels 7 through 13 — 174 to 216 MHz
Channels 14 through 83 — 470 to 890 MHz

Each broadcast station is assigned a specific 6 MHz-wide channel within these bands.

The Video Signal

A television picture does not appear "instantaneously" on the screen of the receiver's cathode ray tube. Instead, it is **scanned** spot-by-spot along 525 horizontal lines from the top to the bottom of the picture by the CRT's electron beam. This process is so rapid that it appears to be instantaneous to the eye. The actual rate is 30 complete pictures or **frames** every second. This rate was determined in part by the bandwidth limitations of the TV system. However, even at a frame rate of 30 Hz, the eye detects some "flicker" when the picture changes. For this reason, the complete frame is divided into 2 **fields** composed of 262-1/2 lines each. These fields are then alternately scanned at a 60 Hz rate.

Figure 7-8 shows how this is accomplished. Field 1 starts in the upper left hand corner and scans across the CRT. It is then deflected back across the tube and starts another horizontal line. All the while, a vertical deflection signal causes each scanned line to be slightly lower on the CRT. After 262-1/2 lines are scanned the vertical signal drives the electron beam back to the top of the picture tube for the start of field 2. Field 2 repeats this process except that its horizontal lines fall exactly in between the lines of field 1. This is called **interlaced** scanning. And, even though each frame is displayed as two separate pictures, the eye sees them as one complete image.

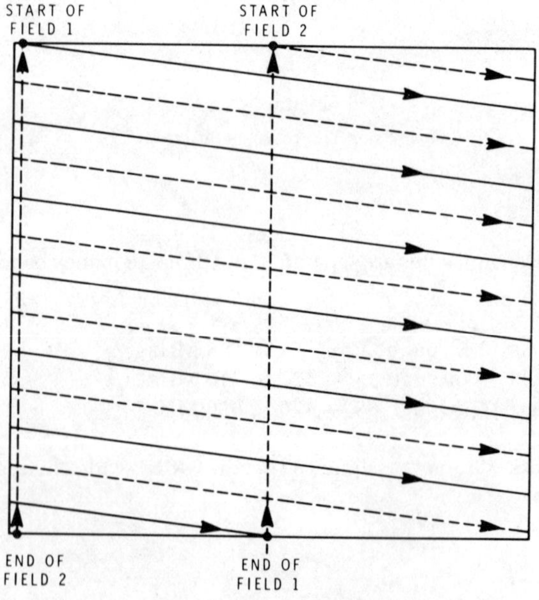

Figure 7-8
Interlaced Scan.

Since the video image is "written" on the CRT using 525 lines, it must also be obtained from the TV camera in the same manner. Figure 7-9 shows a simplified TV camera. The light from the scene being televised is focused onto the light sensitive **target** of the vidicon camera tube. Then the deflection coils cause an electron beam to scan the target and, thus, the video image. The output of the vidicon is a varying voltage which corresponds to the bright and dark portions of the picture. At the receiver, this changing voltage causes the CRT's output to vary in intensity as the picture is scanned. Therefore, it reproduces the camera image on the picture tube screen.

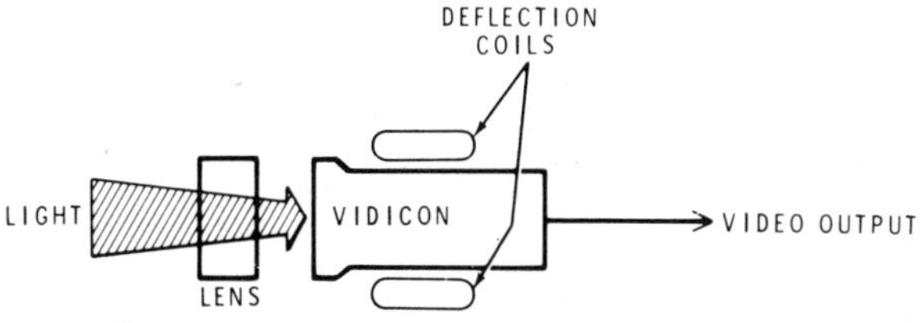

Figure 7-9
TV Camera.

The timing of the scanning at the camera tube and at the receiver's cathode ray tube must be identical — not only in frequency, but also in phase. Therefore, synchronization is necessary. This is accomplished by using special **sync pulses** which trigger the camera deflection circuits.

They are also transmitted as part of the video signal. At the receiver, they are separated and used to trigger the CRT deflection circuits. There are both **horizontal sync** pulses and **vertical sync** pulses. The vertical sync determines the field rate and, therefore, it is sent at a 60 Hz rate. The horizontal sync determines the lines per field. Therefore , it is sent at a 262-1/2 × 60 or 15,750 Hz rate. There is also a **blanking** pulse that cuts off the CRT beam to prevent the sync signals from appearing on the screen.

The composite or complete video signal consists of:

1. The camera video signal,
2. The blanking pulses,
3. The horizontal sync pulses, and
4. The vertical sync pulses.

A composite signal for one horizontal line is shown in Figure 7-10. It is a negative transmission system. This means that the higher the amplitude, the darker the picture detail. At 75% of maximum, the screen is black. Therefore, the sync pulses, which are from 75% to 100%, are effectively blanked out. The reference white level is at 12.5% of maximum with no signal allowed to go lower than this level.

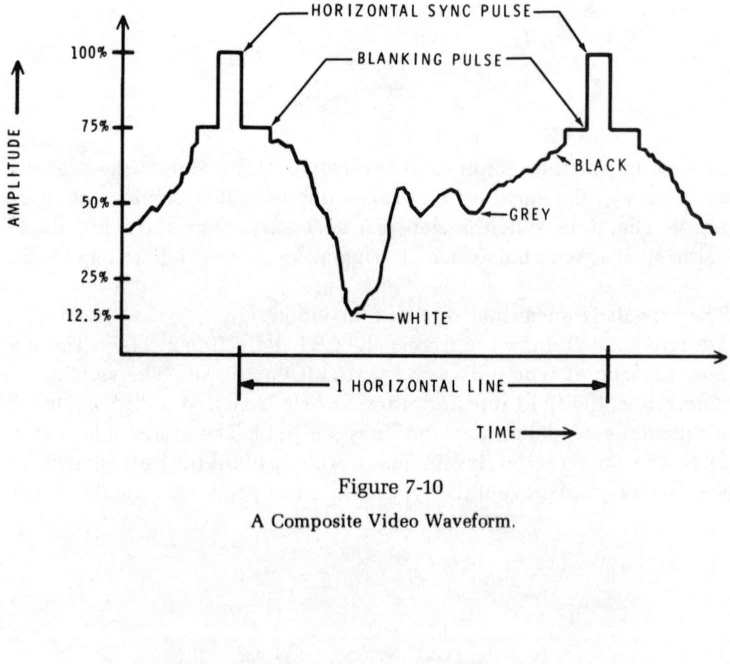

Figure 7-10

A Composite Video Waveform.

The modulated RF wave is shown in Figure 7-11. Since this is a form of AM, the modulation envelope corresponds to the composite video signal.

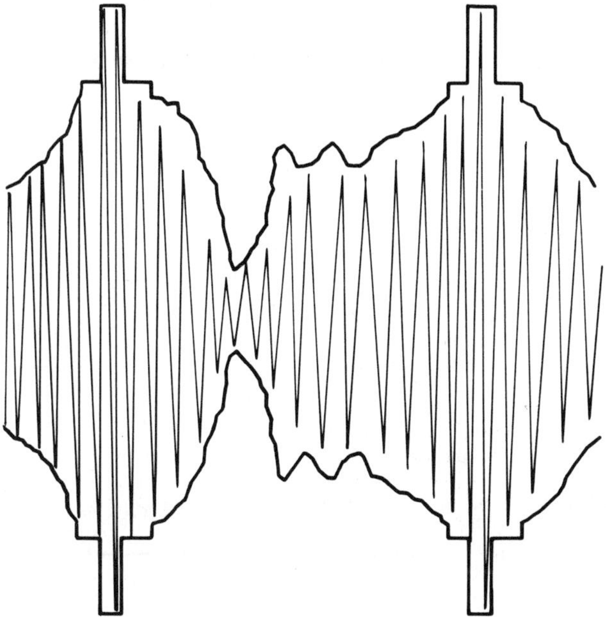

Figure 7-11
A Modulated Video Wave.

The vertical sync pulse occurs at the end of each field. It is shown in Figure 7-12. Note that there are **equalizing pulses** and **serrations** of the vertical sync pulse. These pulses and serrations are used to keep the receiver's horizontal oscillator in synchronization during the long vertical blanking interval. A low pass filter separates the vertical sync pulse and then uses it to synchronize the receiver's vertical oscillator.

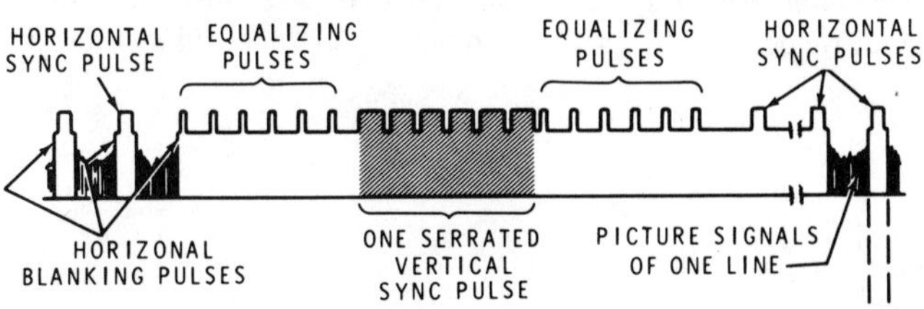

Figure 7-12.
The Vertical Sync Interval.

The TV Transmitter

A greatly simplified block diagram of a TV transmitter is shown in Figure 7-13. It is called an IF modulated transmitter since the video amplitude modulation takes place at an intermediate frequency amplifier. Let's examine this diagram in detail.

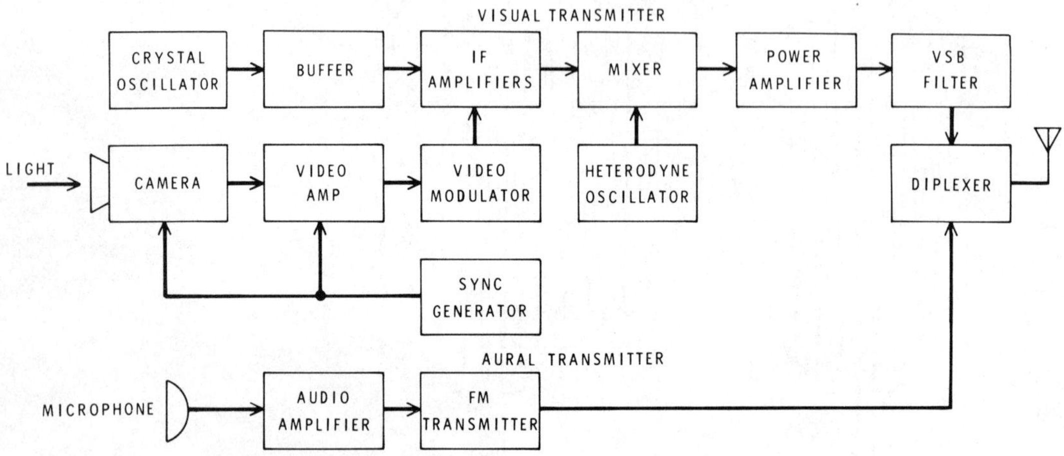

Figure 7-13
Simplified Block Diagram of a TV Transmitter.

The visual transmitter's input comes from a camera and video amplifier. The sync generator supplies scanning voltages and blanking pulses to the camera. It also provides sync pulses to the video amplifier where the composite video signal is formed. The composite video is then applied to the video modulator, which amplitude modulates the cyrstal oscillator's RF signal at the IF amplifier. The IF amplifier increases the amplitude of the modulated RF and also uses a special filter to shape the output frequency response. This is where the vestigial lower sideband is formed.

The mixer and heterodyne oscillator convert the IF signal to the required output channel. The modulated RF is then amplified to the required power level by the linear power amplifier. The VSB filter insures that the vestigial lower sideband is sufficiently attenuated. The visual signal is then applied to the **diplexer.** This device couples both the visual and aural transmitters to the same antenna while preventing either one from coupling to the other transmitter.

The aural transmitter is a straightforward FM broadcast transmitter with a maximum deviation of ± 25 kHz. However, since the aural carrier must be spaced exactly 4.5 MHz from the visual carrier, a special automatic frequency control is used. It effectively "locks" the aural transmitter to the visual carrier by constantly maintaining the 4.5 MHz offset.

The TV Receiver

The simplified block diagram of a TV receiver is shown in Figure 7-14. It is a typical superheterodyne receiver. The RF **tuner** is composed of the RF amplifier, mixer, and local oscillator. It converts the incoming television signal to the intermediate frequency. The standard IF values are 45.75 MHz for the visual carrier and 41.25 MHz for the aural carrier. The frequency inversion occurs at the mixer. The IF amplifiers must be broadbanded to amplify the entire television spectrum. However, their response curve must also be tailored to equalize the lower frequency response with the higher frequency response. The desired IF response curve is shown in Figure 7-15. Such a curve compensates for having two sidebands for video frequencies up to 1.25 MHz and only one sideband for the video frequencies above 1.25 MHz. In short, it eliminates any frequency distortion caused by the vestigial sideband. The result at the diode detector is equal amplitude response over the entire video frequency range.

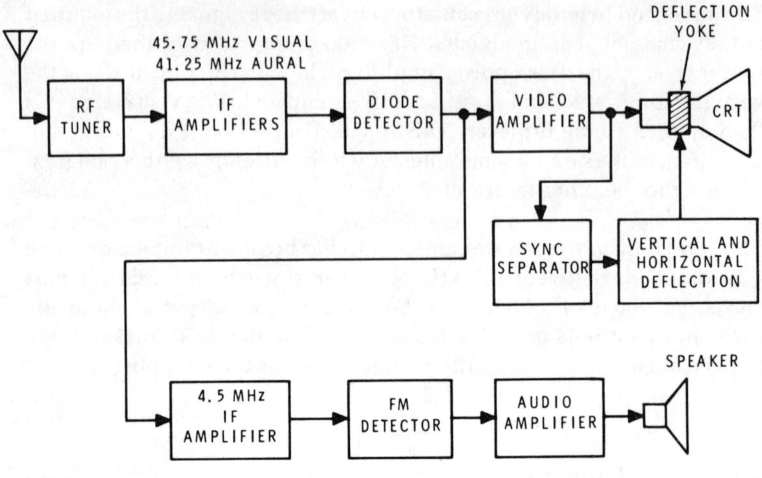

Figure 7-14
Simplified Block Diagram of a TV Receiver.

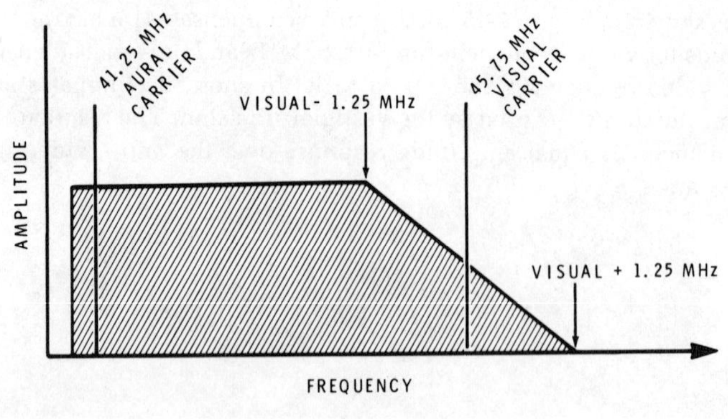

Figure 7-15
IF Frequency Response.

The diode detector not only demodulates the visual AM signal, it also mixes the visual carrier and aural carrier. The difference frequency is 45.75 − 41.25 or 4.5 MHz. This 4.5 MHz FM signal is amplified, detected, and applied to the loudspeaker.

The detected composite video signal is amplified and applied to the CRT and sync separator circuit. At the CRT, it varies the intensity of the electron beam as it scans the screen. The sync separator removes the horizontal sync pulses, equalizing pulses, and the vertical sync pulses from the composite video. These signals are used to synchronize the horizontal and vertical oscillators which, in turn, drive the deflection circuits. The horizontal and vertical deflection circuits drive the CRT deflection yoke and, therefore, cause the electron beam to scan the CRT screen.

DATA COMMUNICATIONS

A data communications system must transmit pulses, that are the output of a data source, from one location to another. An example of this is the transmission of pulses from a teletype unit to a computer, or vice versa. The communication channel itself may be a cable, a switched public telephone line, or even a radio transmission. This section is a brief introduction to the broad and complex field of data communications.

Requirements

The basic requirement of a data communication system is that it must be capable of transmitting rectangular pulses at rates from 100 pps (pulses per second) up to 500,000 pps. However, the most commonly used speeds are 600 to 50,000 pps. Also, the system may be required to transmit in one direction only, alternately in either direction, or simultaneously in both directions. In addition, the error rate must be as low as possible. This requirement is relatively easy to satisfy, provided the communication channel's signal-to-noise ratio is reasonable.

System Types

Basically, there are two types of data communications systems: on-line and off-line. An **on-line system** connects a computer directly to the communications link. On the other hand, the **off-line system** uses an intermediate storage device such as punched paper tape, magnetic tape, or disk to store the data before transmission or upon reception.

An on-line system is generally a **real-time** system. This means that its data transmission requires an immediate response. An example of a real time system is an airline reservation computer. Here, an inquiry from a reservation desk anywhere in the world must be handled and answered rapidly by the central reservation computer. As you can imagine, on-line real-time systems require complex computer software, elaborate communications links, and therefore, are very expensive.

An off-line system is used to transmit data that doesn't require a real-time response or requires no response at all. One example of an off-line system is a national news service such as the Associated Press (AP) or United Press International (UPI). Here, the data, which is news, is transmitted from a central location to the subscribing newspapers, radio, and TV stations. There is no requirement for real-time operation since no response is expected or required.

Modes of Operation

There are basically three modes of operation for data transmission. The first of these is **simplex** which is the simplest mode. It is purely a one-way transmission channel as shown in Figure 7-16A.

The second mode is shown in Figure 7-16B. In this mode, transmission can occur one way or the other but not simultaneously. This is called **half duplex** or **semiduplex** operation.

The third mode, shown in Figure 7-16C, is **full duplex** or, simply, duplex operation. This mode requires two separate communication channels. However, it offers the advantage of simultaneous two-way communication.

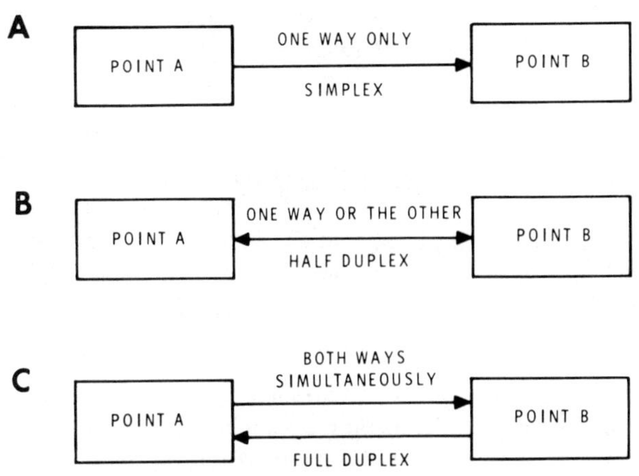

Figure 7-16

Modes of Operation for Data Communication systems.

Modems

Most data communications are over telephone lines, either a switched line dialed by the user, or a special dedicated line. Since DC pulses are not normally sent over telephone lines, a special circuit called a "modem" is used to interface the data equipment to the line.

The term **modem** is a contraction of the words **mo**dulator and **dem**odulator. And, the device does just that — it modulates an audio frequency wave to transmit data and demodulates the incoming audio wave. Thus, the modem consists of an audio frequency transmitter and receiver. It may also convert parallel data to serial for transmission and vice versa on reception.

The modulation technique generally used in modems is frequency shift keying (FSK). In this technique, a specific audio frequency represents an "on" condition, while another frequency represents an "off" condition. The difference between these two frequencies is the frequency **shift**. When data is being transmitted, the pulses "key" or shift the transmitter between the two FSK frequencies just like an FM transmitter. Similarly, the receiver uses an FM detector or phase-locked loop to distinguish between the two frequencies and convert them to DC pulses. An example of an FSK transmission is shown in Figure 7-17.

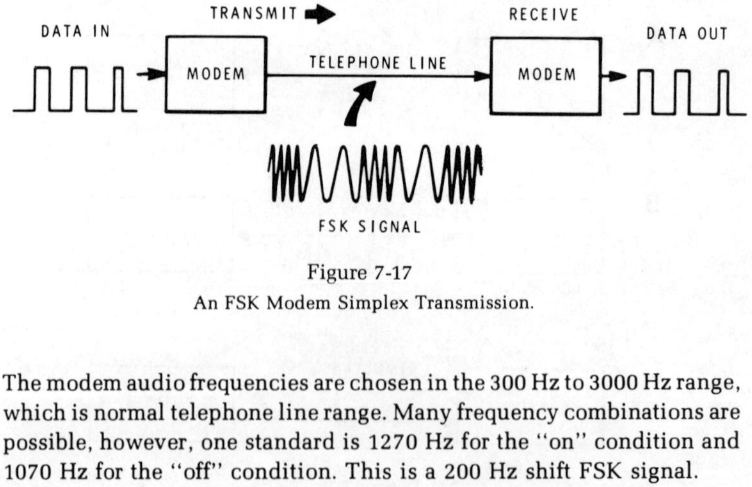

Figure 7-17
An FSK Modem Simplex Transmission.

The modem audio frequencies are chosen in the 300 Hz to 3000 Hz range, which is normal telephone line range. Many frequency combinations are possible, however, one standard is 1270 Hz for the "on" condition and 1070 Hz for the "off" condition. This is a 200 Hz shift FSK signal.

AMERICAN STANDARD CODE FOR INFORMATION INTERCHANGE

	COLUMN	0	1	2	3	4	5	6	7
ROW	BITS 4321 ⇩ 765 ⇨	000	001	010	011	100	101	110	111
0	0000	NUL	DLE	SP	0	@	P	\	p
1	0001	SOH	DC1	!	1	A	Q	a	q
2	0010	STX	DC2	"	2	B	R	b	r
3	0011	ETX	DC3	#	3	C	S	c	s
4	0100	EOT	DC4	$	4	D	T	d	t
5	0101	ENQ	NAK	%	5	E	U	e	u
6	0110	ACK	SYN	&	6	F	V	f	v
7	0111	BEL	ETB	'	7	G	W	g	w
8	1000	BS	CAN	(	8	H	X	h	x
9	1001	HT	EM	)	9	I	Y	i	y
10	1010	LF	SUB	*	:	J	Z	j	z
11	1011	VT	ESC	+	;	K	[	k	{
12	1100	FF	FS	,	<	L	\	l	¦
13	1101	CR	GS	-	=	M	]	m	}
14	1110	SO	RS	.	>	N	^	n	~
15	1111	SI	US	/	?	O	—	o	DEL

Explanation of special control functions in columns 0, 1, 2 and 7.

NUL	*Null*	*DLE*	*Data Link Escape*
SOH	*Start of Heading*	*DCI*	*Device Control 1*
STX	*Start of Text*	*DC2*	*Device Control 2*
ETX	*End of Text*	*DC3*	*Device Control 3*
EOT	*End of Transmission*	*DC4*	*Device Control 4*
ENQ	*Enquiry*	*NAK*	*Negative Acknowledge*
ACK	*Acknowledge*	*SYN*	*Synchronous Idle*
BEL	*Bell (audible signal)*	*ETB*	*End of Transmission Block*
BS	*Backspace*	*CAN*	*Cancel*
HT	*Horizontal Tabulation (punched card skip)*	*EM*	*End of Medium*
LF	*Line Feed*	*SUB*	*Substitute*
VT	*Vertical Tabulation*	*ESC*	*Escape*
FF	*Form Feed*	*FS*	*File Separator*
CR	*Carriage Return*	*GS*	*Group Separator*
SO	*Shift Out*	*RS*	*Record Separator*
SI	*Shift In*	*US*	*Unit Separator*
SP	*Space (blank)*	*DEL*	*Delete*

Figure 7-18
The ASCII Code

Using modems and a single telephone line allows both simplex and semiduplex operation. However, by using a form of frequency division multiplex, full duplex operation can be achieved on a single telephone line. This is done by transmitting in one direction at 1100 to 1300 Hz and in the other direction at 2100 to 2300 Hz. Of course, the modems must be adjusted for this type of operation. But nonetheless, full duplex operation is achieved using a single line.

Data Communications Codes

The most widely used data communications code is the American Standard Code for Information Interchange, abbreviated ASCII. It is a 7-bit binary code that is used in transferring data between computers and their external peripheral devices and in communicating data by radio and telephone lines. Seven bits can represent a total of 2^7 or 128 different states or characters. The ASCII code is used to represent the decimal numbers 0 through 9, the letters of the alphabet including both upper and lower case, plus other special control symbols. The standard ASCII code is shown in Figure 7-18.

The 7-bit ASCII code for each number, letter or control function is made up of a 4-bit group and a 3-bit group. Figure 7-19 shows the arrangement of these two groups and the numbering sequence. The 4-bit group is on the right and bit 1 is the least significant bit. Note how these groups are arranged in rows and columns in Figure 7-18.

Figure 7-19

To determine the ASCII code for a given number, letter, or control operation, you locate that item in the table. Then you use the 3- and 4-bit codes associated with the row and column in which the item is located. For example, the ASCII code for the letter L is 1001100. It is located in column 4, row 12. The most significant 3-bit group is 100, while the least significant 4-bit group is 1100.

There are both 6- and 8-bit special versions of the ASCII code. In addition, the International Business Machines Corporation (IBM) uses another 8-bit coding system called Extended Binary Coded Decimal Interchange Code (EBCDIC) instead of ASCII, for its peripheral and data communications operations.